Manuel Mutende

Study of the feasibility of renewable energy in the rural community

Manuel Mutende

Study of the feasibility of renewable energy in the rural community

Gorongosa District - Sofala

ScienciaScripts

Imprint
Any brand names and product names mentioned in this book are subject to trademark, brand or patent protection and are trademarks or registered trademarks of their respective holders. The use of brand names, product names, common names, trade names, product descriptions etc. even without a particular marking in this work is in no way to be construed to mean that such names may be regarded as unrestricted in respect of trademark and brand protection legislation and could thus be used by anyone.

Cover image: www.ingimage.com

This book is a translation from the original published under ISBN 978-620-6-76210-2.

Publisher:
Sciencia Scripts
is a trademark of
Dodo Books Indian Ocean Ltd. and OmniScriptum S.R.L publishing group

120 High Road, East Finchley, London, N2 9ED, United Kingdom
Str. Armeneasca 28/1, office 1, Chisinau MD-2012, Republic of Moldova, Europe
Printed at: see last page
ISBN: 978-620-8-23769-1

Contents

Dedication

To my mum 'Carolina Manuel Mavie Mutende' and my dad 'Alberto Anguine Mutende' (in memory). Especially my daughter Ayla Manuel Mutende.

Thank you

With all due respect and with the understanding of valuing the religion of the lecturer and of every friend who will have access to my coursework, I would first like to thank the creator of all things, because without HIM nothing would exist, my ALL-POWERFUL GOD "JESUS" for giving me understanding and a lot of spiritual energy to be able to carry out this majestic work, and for the rich and graceful breath he has given me day after day.

I would like to thank my family and my siblings Pedro, Emilia, Deolinda, Jose, Alberto, and Monica, especially my wife Silviana Mujovo Mutende, my brothers-in-law Alfredo Joaquim and Hellencia Alberto, my friends Cristian Coulon, Dinis Simbe, Orlando Guta, Alcamate Dossa, Helton de Sousa, Goncalves Mairoce, Nelson Baptista, Mussa Nohiu, Florencio Chirruque, Meque Batana, Tsambe, Alfoi, Vingan^o, Adji Bacalhau, Adelaide Jemuce, Arne Feijao, Chelsea Jaime Rodrigues Belmiro Salle and the others, and to all those who contributed to my academic training to reach Master's level.

To my fellow students on the Master's programme in Energy Engineering and Management II Building (2018 - 2020), more specifically to my study group from 2018 to 2019, with whom I travelled through a year of intense study.

I would particularly like to thank my mother, Carolina Manuel Mavie Mutende, and my daughter, Ayla Manuel Mutende, for the trust they have placed in me, for giving me strength day after day, and my tutors, Prof. Dr. Eng^{0} . Salvador Carlos Grande and MSc. Eng^{a} . Beatriz Reyes Collado, for instructing me with the aim of completing this degree.

Finally, I would like to thank the Gorongosa District Administrator, Dr Manuel Jamaca, in particular the Mateus Catique village governor, and the leaders of Zambezi University, especially those in my circle of study, the Postgraduate Department of the Faculty of Science and Technology, in particular the Director, Doctor Eng^{0} . Harold Chate.

Thank you very much!

Epigraph

0 man makes his pianos, but it is the LORD who directs the realisation...
Proverbs - 16:9

Summary

This book studies the feasibility of renewable energy in the rural community of Catique, in the district of Gorongosa - Sofala province, in the region of the Vanduzi administrative post, an area outside the PNG reserve. On the one hand, the dissertation is focussed on determining a system for producing sustainable, viable electricity that meets the needs of the Catique community. To complete it, the author used the following methodological tools: a literature review, survey work and data collection using meteonorms software version 7.2 to collect the district's mobile climatological data.

The study was carried out in the belief that renewable energies have the function of moving the community without access to electricity in favour of sustainability, offering residents access to electricity.

The development of this socio-economic system fulfils the government's challenges, in line with the strategy of universal access to electricity for all by 2030. It also aims to forecast energy demand for the next 10 years, taking 2020 as its starting point, and its function is to be a low energy consumption system for the Mateus Catique community.

Thus, the author opts for a solar-photovoltaic proposal, given that the hydroelectric system poses several environmental problems compared to solar, because it is designed to meet the needs of what is most widely applied on a global scale, and consists of high-tech equipment for connecting the system.

Keywords: Renewable Energy, Photovoltaic Energy, Hydroelectric Energy, Sustainability and Energy Viability.

CHAPTER 1

Introduction

1.1. Framework

This work is for a master's degree in Energy Engineering and Management, and the subject falls within the objectives of the course curriculum, and is connected to the following lines of research: Energy Policies, Energy Analysis, Forecasting and Demand, and Energy and the Environment. On the other hand, it fits in with government programmes linked to actions to promote the production of new and renewable energies as a way of reducing energy poverty levels in rural communities and with the objectives of sustainable development to guarantee access to energy for all.

Thus, in today's world, almost two billion people found around urban centres and isolated rural areas have no access to basic energy services. This is the energy paradox that characterises the 20th and 21st centuries. This energy inequality affects two-thirds of Africa's population, which largely depends on traditional biomass for its energy supply (Sebastiao, 2016).

According to Sebastiao (2016), in Mozambique, the number of families with access to electricity is still low, although 99 per cent of the districts already have electricity networks and energy sources installed, more than 80 per cent of the population still does not have access to electricity.

The availability of electricity is one of the key components in the district's economic development. The town is currently supplied with electricity from Cahora Bassa, from the Inchope Administrative Post (Manica province), after its reconstruction, with a capacity of around 33 kVA (ADG, 2017p.5).

One of the most significant challenges for the development of the Mofambican economy is access to electricity for all. The government, through the energy atlas, has shown several studies that suggest Mozambique's renewable energy resources are sufficient to meet the needs of communities in the short and medium term, taking into account the main factors сото demographic growth and economic growth.

Mozambique has the potential to produce electricity from various sources. It has strong and constant winds by the sea and in the high inland regions. They have adequate solar intensity throughout the national territory. All of these renewable sources can be converted into energy (Namburete, 2014).

1.2. Justification

In 2009, the Government of Mozambique approved the New and Renewable Energy Development Policy, establishing the evaluation of new and renewable energy resources as one of the strategic priorities for implementation.

According to the Strategy, the global energy issue has four major challenges to resolve:

- O growing risk of energy supply disruptions from non-renewable sources;
- The threat of environmental degradation from the production and use of fossil fuel energy;
- Energy "poverty" (i.e. the lack of access to modern energy sources by the most economically disadvantaged sections of the population);

- Sustainability (i.e. "the ability to meet the needs of the present without compromising the ability of future generations to meet their energy needs");
- The diversification of the energy matrix.

Thus, the Energy Strategy (EE) aims to prepare the country for the transition to a sustainable energy future by expanding the energy supply matrix, favouring renewable energy sources and simultaneously guaranteeing increased access by larger sections of the population to the benefits of modern energy, in particular electricity, which can be produced with diversified primary energy sources included in the supply matrix, given the proven relationship between access to electricity and human development.

The same Strategy calls for the acceleration of electrification efforts, prioritising rural areas through the expansion of the National Energy Transmission Network (RNT) and alternative energies, including through the use of low-cost solutions and strengthening collaboration between institutes such as Electricidade de Mozambique (EDM) and Fundo de Energia (FUNAE), as well as the introduction, in the investment packages, of a percentage aimed at financing electrical equipment and goods designed to encourage the productive and efficient use of energy (low-consumption, high-efficiency lamps).

On the other hand, there is an ambitious target envisaged or set out in the 2030 Sustainable Development Goals (SDGs) that defines, among other things, global coverage in the provision of efficient and quality energy. In addition to expanding infrastructure and modernising technology for the provision of sustainable energy services.

It is with these global and national objectives in mind that projects are designed to align with and implement this policy, on the one hand, but on the other, due to the diversity of the country's geographical regions, studies continue to be carried out in order to realistically find the most appropriate responses for each rural community.

This project therefore aims to study and characterise the viability of the energy potential in the Mateus Catique community in Gorongosa, in order to insert a viable electricity generation system for the community.

1.3. Objectives

1.3.1 General Objective

The overall aim of this work is to determine a sustainable, viable system for producing electricity that meets the needs of the Mateus Catique community in Gorongosa - Sofala.

1.3.2 Specific objectives

- Verify the state of the art of renewable energy systems for generating electricity in the world and in Mozambique;
- Identify the energy potential of the main renewable energy sources in the Mateus Catique community;
- To study and design an efficient and sustainable renewable energy system for the Mateus Catique community in Gorongosa - Sofala;
- Suggest projects to meet energy needs by 2030;

1.4. Work structure

This dissertation is divided into 5 chapters, described below:

Chapter 1, which consists of the introduction, presents the framework of the research, the relevance, the justifications, the problematisation, the objectives and the structure of the work;

0 Chapter 2 presents the literature review, which covers concepts relating to the types of renewable energy sources that exist, the region's climatic conditions, spatial organisation and materials used on the African continent and in Mozambique in concrete terms;
Chapter 3 presents the procedures used to collect data at Gorongosa district level, as well as in relation to the community of Catique, and is also dedicated to the interpretation of the main concepts used in the work, as well as the calculation sheet for the sizing of the most abundant renewable energy system in Gorongosa;
0 Chapter 4 presents the results obtained and relevant to the research, as a result of the bibliographical review, and the survey work, through the sizing that took place in that chapter, as well as a sequence of steps for the correct choice of reliable renewable energy system to meet the population demand of the Mateus Catique community.
Chapter 5 presents the final considerations, which set out the general conclusions and offer some reflections on the possibilities for future developments in this field.

2. Renewable energy sources

2.1. Energy and Renewable Energy

The word energy derives from the Greek *svspysia, (energy),* which means work and was first used by Aristotle. In the modern sense, this concept was created in association with thermodynamics in the mid-19th century and used to describe a wide variety of phenomena.

The usual definition states that *"energy is the measure of the capacity to do work".* However, this definition is not entirely correct and only applies to certain types of energy that are fully convertible into other forms. In 1872, Maxwell proposed a definition seen as the most correct, *"energy is that which allows a change in the configuration of a system, as opposed to a force that resists this change"* (Fortes et al., 2020).

According to Fortes et al. (2020), all concepts present energy as an intrinsic phenomenon of matter, which can manifest itself in various forms (Figure 2.1), with the possibility of inter-conversion, conservation and transformation into mass and vice versa.

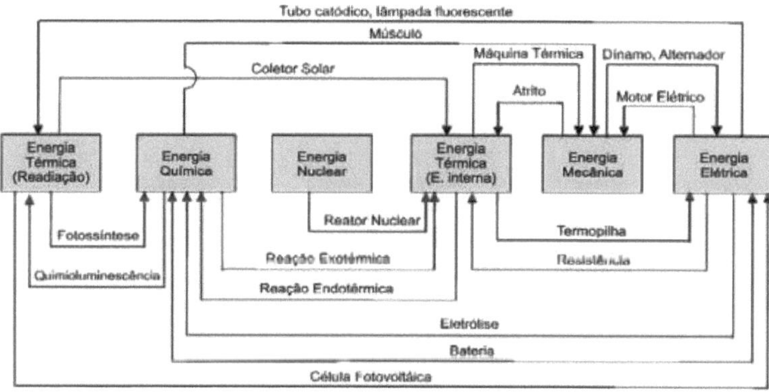

Figure 2.1 - Energy conversion processes (Fortes et al., 2020 p.10)

Renewable energies are those that have replenishment rates equivalent to their utilisation, and this replenishment can occur naturally or artificially (Elliot, 2000). Renewable energies therefore have a number of advantages (see Table 2.1 - The positive and negative aspects of renewable energies).

According to Da Silva (2011), estimates indicate that Mozambique has between 60 and 100 regions with possible characteristics for harnessing mini-hydroelectric plants, wind turbines, photovoltaic generators and others, with only studies needed to show the real potential of these sources.

Examples of naturally renewable sources include solar, wind, hydro and natural biomass. Artificially renewable sources are represented by planted biomass and the waste generated in industries and other processes controlled by humans, including rubbish.

2.2. The Impact of Renewable Energy on Sustainable Development

According to Fortes *et. al.* (2020), renewable energy in Mozambique has a significant impact on the economies of various communities, considering availability, generation, use and future trends.

According to the author, renewable energy sources are increasingly being developed and constitute a significant part of the national energy matrix. The Mozambican territory has characteristics that make it a potential region for energy production, in quantities sufficient for self-sufficiency, a fundamental factor in national socio-economic development and socio-environmental sustainability.

Changes in the local energy matrix should be seen as both a challenge and an opportunity for technological innovation, with a view to expanding access to energy, improving energy security and efficiency, reducing current levels of energy poverty, reducing dependence on fossil fuels and the socio-economic integration of rural and suburban populations.

O Table 2.1 summarises the main aspects (positive and negative) that determine your choice when it comes to deployment.

Table 2. 1 - Summary of the positive and negative aspects of the main renewable energy sources in Gorongosa, Sofala

Energy	Source	Positive aspects	Negative aspects
Hydroelectric	Water	- Low operating costs; - Long-lasting plants; - Numerous jobs are generated during construction; - High efficiency; - Long lifespan; - It does not emit greenhouse gases [GHG] when generating energy.	- High construction costs - Loss of biodiversity; - Removal of native people; - Biological, physical and chemical disorders; - Changes in rainfall have a direct impact on electricity generation.
Solar	Solar radiation	- It comes from a renewable resource; - Proximity between generation and consumption; - Clean, cheap energy; - Small and large-scale installation possible; - Systems require little maintenance.	- Solar panel production process generates GHG; - High cost of solar panels; - Intermittent fountain with efficiency depending on climate variation; - Forms of storage are still not very efficient.
Wind	Wind	- An inexhaustible source of energy; - Low maintenance and high efficiency; - It emits no GHG and generates no waste; - 0 soil and used for other purposes; - Cheap sources of energy.	- Constant winds are needed; - Sound and visual impact - Impact on local birds; - Intermittency.
Biomass	Organic matter	- Carbon neutral;	- Modification

		- Abundant domestic energy source; - Low raw material costs - Waste becomes an input for another process.	local ecosystem; - It can be affected by changes in cultivation regimes; - Lower calorific value; - Difficulty storage.

Source: Fortes et al. (2020).

2.3. Diversification of the energy matrix

2.3.1. Hydropower

Hydric energy is the energy obtained from the potential energy of a body of water. The form in which it manifests itself in nature and in water flows is rivers and lakes and can be harnessed by means of a drop or fall of water (Franca, 2001).

Hydroelectric energy is supplied by hydroelectric systems, which are a set of devices that take the hydraulic energy from waterfalls accumulated in rivers or lakes and transform it into mechanical energy to move a hydraulic turbine (Franca, 2001).

Hydroelectric systems are made up of complex infrastructures that include dams, hydraulic channels, pipelines, turbines and generators.

2.3.1.1. Technical criteria for turbine selection

The selection of the turbine for a given hydroelectric system is fundamental from a technical point of view in order to achieve good performance in terms of yield, without forgetting the economic point of view (Joao, 2016). The following techniques should therefore be taken into account:

- **Drop and Flow**

Firstly, if you only consider the criterion of the height of the useful drop, you can roughly relate the type of turbine to the height of the drop (**Annex 5** - Field of application of hydraulic turbines).

- **Specific speed**

According to Munhoz (2015), knowing the drop and o flow that characterise o exploitation and assuming that the turbine is directly coupled to the generator at a certain rotation speed, the value of the specific speed must fall within one of the ranges shown in table 2.1 below.

Table 2.1 - Specific speed range

Turbine name	Specific speed
Pelton	**$1000 < n < 1800$**
Francis	**$80 < n < 1000$**
Kaplan	**$50<n<80$**
Michell Benki	**$10<n<50$**

Source: Munson, 2002.

2.3.2. Solar Energy

Solar energy is an abundant renewable resource throughout the Earth. Every hour, the amount of solar energy that reaches the Earth's surface is enough to supply humanity for a year (Elliot, 2000).

Solar energy can be used in photovoltaic modules to produce electricity and in solar collectors to heat water or other thermal fluids.

According to Fortes et al. (2020 p. 12) there are basically three ways of capturing and converting solar energy:

(I) or photo-biochemistry, where certain organisms synthesise carbohydrates from water and CO_2 , absorbing solar energy and storing it in chemical bonds;

(II) Electrical or photoelectric radiation occurs through the emission of electrons from the surface of semiconductor materials exposed to high-frequency electromagnetic radiation (i.e. light);

(III)thermal or thermoelectric, through the absorption of radiant energy by a black surface. Diffusion, absorption of photons, acceleration of electrons and multiple collisions occur in the absorbing material, transforming the radiant energy into heat.

2.3.2.1 Comparative study of solar radiation in the world and in Mozambique

Photovoltaic energy has existed for more than 100 years, and today it is used to generate electricity for thousands of homes and industries around the world, and сото Mozambique, since the 1990s this technology has been studied by different authors, with the aim of accelerating the transition to the use of sustainable energies and transforming the lives of inhabitants in different contexts, such сото: o access to lighting at night, whether in schools or health centres сото primary sectors and as a source of alternative energy, among others.

According to Namburete, (2014) Mozambique has a good solar resource, consistent throughout the territory and stable throughout the year, solar radiation on a global scale that varies essentially depending on the atmosphere, geometry and the movement of the planet in relation to the sun, while on a local scale, the variation in solar radiation is mostly associated with the morphology of the terrain, i.e. variations in elevation, slope, exposure and shading.

Mozambique has a high global horizontal piano radiation when compared to other locations in Europe and Asia.

2.3.2.2 Main components of the photovoltaic generator

Photovoltaic module

According to Sousa (2015), the photovoltaic cell is the basic element of the photovoltaic module. It is in the cell that the sun's radiant energy is converted into electrical energy. Photovoltaic modules are electrically connected to each other (in series or parallel) and function as a single generator of electricity.

Figure 2.2 shows the operating diagram adapted for the Mateus Catique community's PT vs photovoltaic generator.

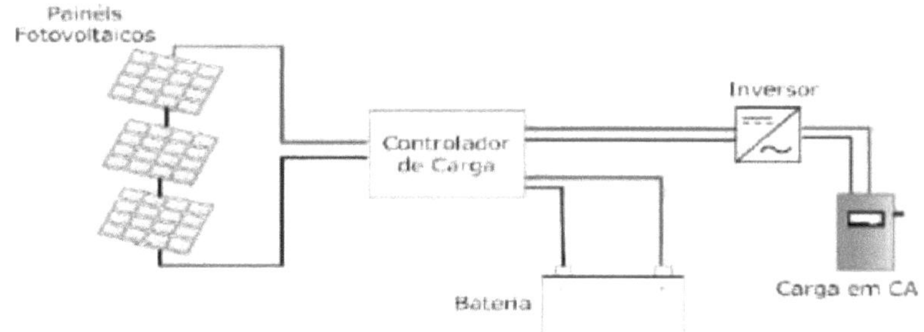

Figure 2.2- Operating diagram adapted for PT vs. generator generator
(Adapted: Scheibler, 2015 apud Santos, 2010).

When connected in series, the voltages are added together and the current remains unchanged. When connected in parallel, the voltages in the cells are equal and the currents are added together.

- **DC-AC inverters**

This is an electronic device whose function is to convert the direct current from the photovoltaic generator into alternating current (Rocha, 2010). In these types of photovoltaic systems it is common to use an inverter connected to the photovoltaic generator.

- **Types of inverters used in photovoltaic systems**

Micro-inverters: this is a more recent technology, which makes it possible for an SFV to be made up of small inverters placed in each individual panel.

Typically, these inverters have a 25-30 year warranty and are 5% to 10% more expensive than *string* inverters, being operated at lower voltages than *strings* (Bernal, 2014).

***String* inverters**: these are connected in series and in lines with the photovoltaic panel, and there can be associates in parallel to the *string*, with a guarantee of usually 10 years. Thus, the loss of energy in voltage drops is greater for AC currents than DC currents and in *strings* the conversion only takes place at the end (Bernal, 2014).

- **O role of batteries in a photovoltaic system**

Batteries or battery banks play an essential role in an isolated photovoltaic system, as they store electrical energy in potential chemical energy and can then convert this energy back into electrical energy (Scheibler, 2015).

Batteries are made up of cells that are the basic units of their constitution, which are responsible for accumulating energy (Santos, 2010).

2.3.3. Wind energy

Wind energy comes from the Latin word aeolicus, belonging or relating to Aeolus, the god of the winds in Greek mythology, and therefore belonging or relating to the wind (Patel, 2006).

Since ancient times, wind energy has been used to power boats propelled by sails or to run the gears of windmills by moving their blades. In windmills, wind energy was transformed into mechanical energy, used to grind grain or pump water.

2.3.3.1. Comparative study of wind resources in the centre, south and north of Mozambique

A study carried out by Pinto (2008), between 1999 and 2003, using the RegCM3 atmospheric model, showed an expectation of significant wind resources in the Mozambique region, with maximum average wind intensities of between 7.0 and 9.0 m/s, between June and August, occurring in most of the coastal regions and highlands of Mozambique.

The use of wind energy in Mozambique is not yet a sustainable alternative due to the high costs of its technology, but it is still a very promising renewable matrix for regions where the conventional electricity grid is not economically accessible to meet the country's future energy demand, since there is currently a reduction in electricity production, associated with the decrease in the flow of the Zambezi River, due to the high rate of evapotranspiration induced by climate change (FUNAE, 2009).

2.3.3.2. Main elements of a wind generator

According to Oliveira *et. al.* (2003) and Patel (2006) the following elements can be distinguished:

- Anemometer: measures the intensity and speed of the wind. It works on average every ten minutes;
- Biruta (direction sensor): captures the direction of the wind. The direction of the wind should always be perpendicular to the tower for best utilisation;
- Pas: they capture the wind, converting its power to the centre of the rotor;
- Generator: item that converts the mechanical energy of the shaft into electrical energy;
- Control mechanisms: adjusting the nominal power to the wind speed that occurs most frequently during a given period;
- Gearbox: responsible for transmitting mechanical energy from the rotor shaft to the generator shaft;
- Rotor: assembly connected to a shaft that transmits the rotation of the blades to the generator;
- Nacelle: compartment installed at the top of the tower consisting of: gearbox, brakes, clutch, bearings, electronic control and hydraulic system;
- Tower: element that supports the rotor and nacelle at the appropriate height for operation. The tower is a high-cost item for the system.

Thus, the force of the wind rotates the blades of a turbine and is thus recovered in the form of mechanical energy in the rotor shaft, which in turn is converted into electrical energy by the generator.

2.3.4. Biomass energy

Biomass - organic material of biological origin, including biodegradable fractions of agricultural and forestry products, including agricultural and animal waste, as well as biodegradable municipal waste (Ministry of Energy, 2013);

According to the Ministry of Energy (2013), woody biomass - includes all organic material from trees or shrubs (branches, bark, roots). Biomass energy - is energy derived from the combustion, fermentation or transesterification of organic materials

(firewood, charcoal, pomace, biodiesel, ethanol, biogas and others).
Biomass can be used widely, directly or indirectly. The lower percentage of global and localised atmospheric pollution, the stability of the carbene cycle and the greater use of labour can be mentioned as some of the benefits of its use.
Likewise, in relation to other forms of renewable energy, biomass, like chemical energy, has a prominent position due to its high energy density and the ease with which it can be stored, exchanged and transported.

2.3.4.1. Comparative study of the use of biomass in the world and in Mozambique

According to the Ministry of Energy, (2013) Mozambique has enormous biomass resources that are still under-exploited in a sustainable way, including firewood, charcoal, agro-forestry residues, municipal organic waste and other resources.
Mozambique has benefited from three national inventories, which reflect the country's biomass potential. Around 65.3 million hectares constitute forest biomass, of which 1.7 billion cubic metres are woody formations and 1.1 billion tonnes are wood of various commercial and non-commercial species.
It should be noted that favourable agro-ecological conditions give the country the potential to produce 22 million tonnes of above-ground woody biomass, which constitutes an annual growth rate of 2% o equivalent to 0.4-1.6 m^3 /year/ha.
However, it should be borne in mind that not all of the potential referred to is intended for the production of firewood and charcoal, as biomass is not limited to the living biomass that has usually been inventoried in the country.
Plant biomass goes beyond the quantities estimated in inventories, considering that o volume is estimated on the basis of established forest inventory methods, and these are orientated towards o trunk volume, neglecting dead biomass (forest waste), as well as сото o branch volume, o which occupies around 50% of the total tree volume.

Factors affecting the supply of wood fuels in Mozambique

The supply of wood fuels is influenced by the current stage of the resource, management and accessibility of the resource, and the influence of these factors is seen in an exponential positive or negative way, according to the following approaches:

a) Current stage of the appeal

The current forest cover corresponds to 51 per cent of the total forest and gives the country a wealth of vegetation, especially in the provinces of Zambezia, Sofala, Niassa, Cabo Delgado, Inhambane and Nampula.

b) Resource accessibility

In the country, access to wood fuels is not a limiting factor, particularly in rural areas where the population depends on these resources to meet domestic energy needs. However, in urban and peri-urban areas, access has been reduced due to high market prices.

c) Resource management

After independence, various projects to establish energy plants were developed in the country's largest urban centres, most notably in the provinces of Maputo, Beira and Nampula, but the results were unsatisfactory.

2.4. Legal Framework for Renewable Energy Sources

0 Mozambique's energy sector has adopted a number of strategies and policies with the aim of promoting and facilitating access to renewable energies, as well as expanding the coverage of the National Electricity Network (REN), improving technological efficiency, increasing the availability of electricity at competitive prices, promoting private sector participation and environmentally sustainable practices. Thus, solar and hydropower are undoubtedly the predominant resources in the Gorongosa region. And hydropower is the main source of electricity generation, and there are plans to develop the national hydroelectric park.

In the general context of this work, it can be seen that renewable sources can also contribute to electricity generation and increase national and regional electricity security. Mozambique has enormous energy resources derived from biomass, the exploitation of which is not yet sustainable. However, several efforts have been made over time to reverse this situation, and some instruments have already been drawn up that are worth highlighting, namely: The Energy Policy approved by Resolution 5/98 of 3 March, the New and Renewable Energy Development Policy approved by Resolution 62/2009 of 14 October, the Forestry and Wildlife Law, approved by Law 10/99 of 7 July, the Land Law approved by Law 19/97 of 1 October, are important instruments for the sustainable development of renewable energy resources in Mozambique.

2.5. Rural Electricity and the Contribution of Renewable Energies

Rural electrification is an instrument of energy policy used in many developing countries and is one of the ways found to minimise the shortage or *deficit of* energy supplies in rural areas (Sebastiao, 2016).

One of the aims of electrification is to eradicate poverty, alleviate social problems and boost people's priorities, and it must be part of a rural development strategy.

The method of supplying electricity to these areas varies and can include isolated generators serving collective or individual consumers. However, the variation of these methods is dependent on local circumstances and the degree of saturation of the electricity supply.

However, this concept faces some limitations in its applicability associated with issues of a cultural, economic and social nature. In reality, the inhabitants of these areas are often culturally unprepared to receive this type of technology, or are unable to afford it.

The objectives, planning, realisation and operation of rural electricity projects cannot be separated from problems such as poverty, concerns about environmental degradation, rural development and energy needs in general (Forjaz, 2011).

Forjaz (2011) and Sebastiao (2016) illustrate the following evidence of electrification and the development of rural areas:

- In the closing decades of the 19th century, before electrification, a significant number of measures were implemented to encourage the development of rural regions. Rural electrification stimulates the development of local industry so that rural areas are prepared for other stages of development.
- On many farms in Ireland, electricity led to increased productivity and reduced costs. The availability of electricity allowed industrial activities to advance, and led to the development and establishment of service supply businesses and the manufacture of

equipment and tools.

- In Ireland, rural development would have been further advanced with a well-planned integrated rural action programme. Electricity is a process, which when implemented effectively, contributes to harmonious development.

In this sense, improving living conditions in rural areas can be done by satisfying basic needs and by promoting small-scale industrial activities (introduction of mills) aimed at increasing economic independence.

2.6. Literature Review

According to Sebastiao (2016), he concludes in his work on the Mozambique electrification model: the importance for development that:

- It has been proven that electrification brings benefits to rural communities, the economy, institutes and the environment.
- Without sufficient and adequate energy resources, developing countries are unable to promote the social and economic development that is crucial for sustained growth.
- The research methodology based on the case study allowed us to take a deeper look at the subject in question. The qualitative nature of this methodology facilitates the analysis of social and business contexts, behaviours and characteristics. The main objective is to analyse the potential of implementing new and renewable energies in rural areas and whether investing in this type of energy in Mozambique is an economically, socially and environmentally sustainable alternative.
- Considering the SWOT analysis and specifically addressing the weak points, they also conclude that one of the main objectives of rural electrification is fundamentally the fight against poverty.

According to Joao (2016), in his work on the economic feasibility study of mini-hydro and *diesel* power plant projects for the town of Majaua, in the province of Zambezia (Mozambique), he concludes that:

- Some of the parameters studied in this work were considered to be unvarying over the operating period of the power stations (annual gross and net revenue, average annual cost of energy produced, energy sales tariff, annual utilisation of installed power, fuel costs, etc.), which greatly distorted the economic profitability results obtained, which are the subject of this study.
- The consultant should advise the Mozambican authorities to invest in the Mini-hydroelectric Power Station Project.

According to De Castro *et al.,* (2009) in their paper entitled: The Importance of Alternative and Renewable Sources in the Evolution of the Brazilian Electricity Matrix, presented at the V Seminar on Generation and Sustainable Development organised by Fundacion MAPFRE, concludes that:

- Faced with the inability of reservoirs to regularise generation on an annual basis, it is necessary to supplement generation on a seasonal basis, using sources with such a vocation.
- To make the matrix defined in the planning feasible, a first solution would be to replace General Auctions with Source Auctions. Matrix exposed to A-5 and A-3 auctions.
- However, more than just a policy to promote renewables, there needs to be planning of the matrix that takes into account safety, economic and environmental sustainability variables.

According to FUNAE (2019), in the manual entitled Renewable energy project portfolios - water and solar resources - 2-. Edifao, concludes that:

- 0access to information via radio, mobile phones or television is an important way of creating economic, civic and educational opportunities.
- Using energy for lighting allows children and adults to study at night and improves safety, an issue of particular importance for women and girls.
- The use of energy for charging mobile phones, supplying machinery, lighting and refrigeration can be crucial for the development of small businesses.
- Energy also saves time for cooking, cleaning and fetching water, which are time-consuming tasks traditionally carried out by women in Mozambique. Especially in rural areas, women have limited access to financial resources and often encounter major barriers to starting small businesses. Therefore, women can benefit enormously from improved conditions for setting up small businesses, for example with greater access to energy.
- 0 access to affordable energy and therefore an essential input for the economic development of women in rural communities and in general.

CHAPTER 3

3. Case Study: The survey work

This chapter makes a detailed diagnosis of the location of Mateus Catique village, Mozambique - Gorongosa, as well as the sizing of the most abundant and favourable systems for this locality, the photovoltaic and hydroelectric systems to be implemented in the community in the District of Gorongosa.

Thus, the results of this chapter will be obtained from the definition of the case study and the visit to the Mateus Catique village/community in Gorongosa.

3.1. Mozambique's situation

In Mozambique, 69% of the inhabitants live in rural areas, according to projections made in 2011 by the United Nations (*Mendes, 2018*). Graph 3.1 shows this data.

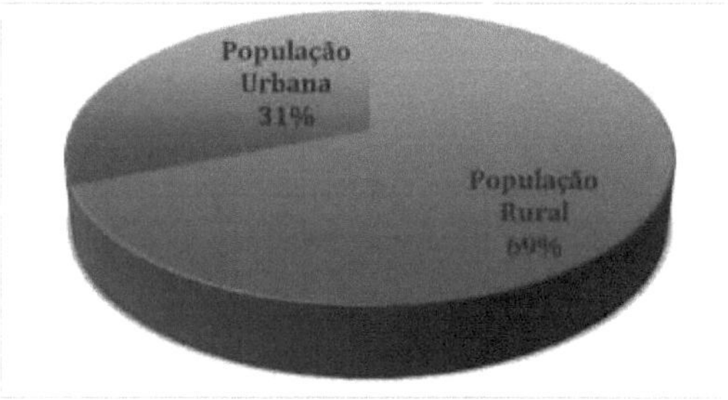

Graph 3. 1 - Percentage of Mozambique's population living in urban and rural areas areas (adapted from: Mendes, 2018 apud United Nations, 2012 p. 156-157)

According to data from the INE (2017), there were around 27 million inhabitants in 2017, compared to 16 million in 1997, which shows a demographic growth of around 64 per cent from 1997 to 2016.

The population, we can see, is relatively evenly distributed across most of the provinces, with the exception of Zambezia and Nampula, which are the most populous provinces. INE's demographic indicators show that the country's population is quite young, with more than half of the population (64.5%) aged under 25, and only 3% aged 65 or over.

The educational level is still low, and is worse in rural areas. The literacy rate stands at 55 per cent, and is higher in the younger age groups and among men. The population is predominantly rural and the vast majority of the labour force (75.2% in 2007) is involved in agriculture, livestock, hunting, fishing and forestry (ADG, 2017).

In Mozambique, for example, more than 80 per cent of the population does not have access to electricity. This problem is a consequence of the inability to expand the electricity network to certain more remote and isolated areas. On the other hand, it is very costly for the government to provide access to electricity in these areas when there is no prospect of a return on the capital invested, since the majority of the inhabitants are poor (Ministerio de Energia de Mozambique (MEM) & Fundo Nacional de Energia de Mozambique (FUNAE), 2009).

In the same context, only 10.5 per cent of households have access to electricity, of which more than half are in Maputo and the surrounding areas. All the provincial capitals and most of the municipalities are also supplied with electricity, сото о case of Sofala Province - Gorongosa District. Most of these urban centres are connected to the national electricity grid, which is owned and operated by the power utility Electricidade de Mozambique (EDM) (Sebastiao, 2016).

Most of the population is concentrated in a small number of urban centres. Increasing access to electricity in these areas has proved difficult and expensive (Ministry of Energy, 2013).

On the one hand, Mozambique receives a considerable amount of sunshine, with an average annual radiation of 5 kWh / m^2 / day, which offers very favourable conditions for solar photovoltaic and thermal energy development. And due to its location near the coast, it offers optimum conditions for the development of hydroelectric and/or mini-hydro energy (UNESCO, 2012).

3.1.1. Main energy matrix in Gorongosa

3.1.1.1. Hydropower

Hydroelectric energy comes from the condensation, precipitation and evaporation of water, factors caused by solar irradiation and gravitational attraction. Hydroelectric power stations have the capacity to transform the kinetic energy and gravitational potential of a large volume of river water into mechanical energy (in the turbine) and electrical energy (in the generator) (Fortes et al., 2020).

Favourable locations for the installation of hydroelectric power stations are high drop-off points, on steeply sloping streams, formed by rapids or waterfalls. In order to produce electricity, it is necessary to unify the slopes by building dams or reservoirs that interrupt the normal course of the river (Fortes et al., 2020).

Mozambique has always invested and will continue to invest in harnessing its renewable energy potential to bring more and better quality energy to more and more Mozambicans. Hidroelectrica de Cahora Bassa (HCB) was created with the exclusive objective of exploiting, under concession, the hydroelectric exploitation of Cahora Bassa and, in general, the production, transport and commercialisation of electricity from the Cahora Bassa dam, including its import and export, and may carry out all acts related to its object, necessary or useful for its realisation.

HCB is one of the largest hydroelectric dams on the African continent, with a plant equipped with 5 turbines of 415MW each, producing a total of 2075MW, on the Zambezi River in the province of Tete. It is surpassed only by the Assuao dam in Egypt in terms of the size of its reservoir. On 27 November 2008, HCB reverted to the Mofambican state, which now owns 85% of the company's share capital.

The potential energy accumulated at high altitudes is transformed into kinetic energy as the speed increases along the gradient, and is also dissipated on its way down through rivers, bends and obstacles, finally becoming heat for the environment. What hydroelectric generation does is intercept this descending water and convert its hydraulic energy into electrical energy through hydro-generators.

3.1.1.2. Comparative study of water resources in the world and in Mozambique

According to Namburete, (2014) Mozambique has a good hydroelectric resource, о тара of hydroelectric potential shows that the provinces of Sofala, Zambezia and Niassa are the areas with the greatest energy potential due to the combination of the most favourable flow and terrain morphology, identifying places of high drop. However, the greatest potential for production lies along the Zambezi River, where the highest flows are recorded.

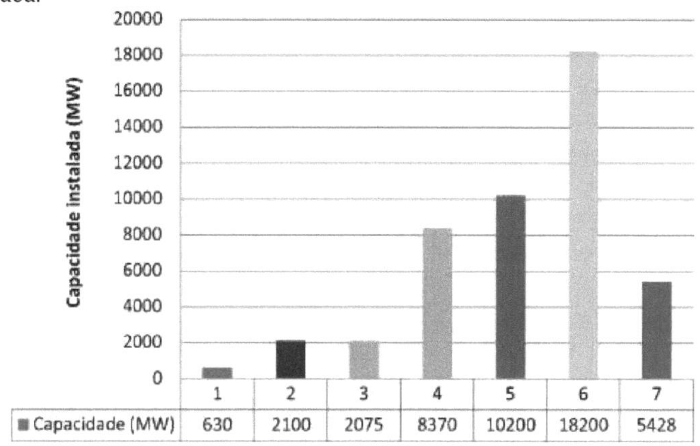

Graph 3.2 - Comparison of water resource capacity in the world and in Mozambique
(Adapted from Namburete, 2014)

Table 3. 1 - Comparison of water resources in the world and in Mozambique

Parents	Rio Local	Capacity (MW)	Dam Name
Portugal	Lima	630	Alto do Lindoso
Egypt	Nile	2100	Upper Assua Dam
Mozambique	Zambezi	2075	Cahora Bassa
Brazil	Tocantins	8370	Tucurui I and II
Venezuela	Caroni	10200	Guri or Simon Bolivar
China	Yang Tse	18200	Three Gorges
Canada	Churchill Falls	5428	Churchill Falls

Source: Adapted from Namburete (2014).

Mozambique has a high potential compared to good locations in Europe and Africa, as seen in Graph 3.2 and Table 3.1 above, which shows that the Cahora Bassa dam is currently the largest producer of electricity in Mozambique, with a capacity of over 2000 megawatts, supplying Mozambique (close to 250MW), South Africa (1100MW) and Zimbabwe (400MW). And these days, according to FUNAE (2019), negotiations are underway to supply Malawi with electricity from Cahora Bassa.

3.2. Techniques used for data collection

This work will have several stages, which will be defined according to the tasks to be carried out and the respective methodologies to be applied. The following stages have been defined:

Bibliographic review: analysis of methods for constructive and mechanical characterisation of solutions Through bibliographic research on new energy matrices for hydroelectric, photovoltaic, wind and biomass systems in Mozambique and other

African countries with similar conditions and climates, it should be noted that the bibliography in this area, specifically in Mozambique, is limited.

Visit to the Mateus Catique **community**, and survey of solutions through photographic recording, visual observation, measurements, and others that may be necessary. In order to carry out a systems analysis and taking into account the socio-economic conditions of the inhabitants, the climate and the materials available;

Data collection: the climatological data for the district was analysed and made available by the METEONORM 7.2 tool. The version can be *downloaded* from Google, **https://meteonorm.software.informer.eom/7.2.**

The data relating to population growth, GDP and CPI at provincial and district level was obtained from INE, Sofala delegation, Cidade da Beira - or you can go to **https://www.ine.gov.mz/** pages of INE, Mozambique.

3.3. Characterisation of the study site

The area of the district is 6,722 km^2 and its population is estimated at 181,000 inhabitants (INE, 2017). The Mateus Catique community is located in the province of Sofala, in the district of Gorongosa, in the administrative post of Gorongosa, as shown in figure 3.1. and annex 1. The region forms the natural boundary of the extreme east of the district with the districts of Muanza and Cheringoma, and simultaneously serves as the eastern boundary with the Gorongosa National Park reserves, to the north with the district of Maringue, to the south with the districts of Nhamantanda and Dondo (ADG, 2017).

The Mateus Catique community, as the residents call it, is the most populous village in the Vanduzi region, compared to the communities of Vinhos, Casa Banana and Nhamacola (in PNG). It is one of the villages with the best conditions in terms of urbanisation and local sanitation units, although the people still sometimes have a culture of open-air faecalism at night.

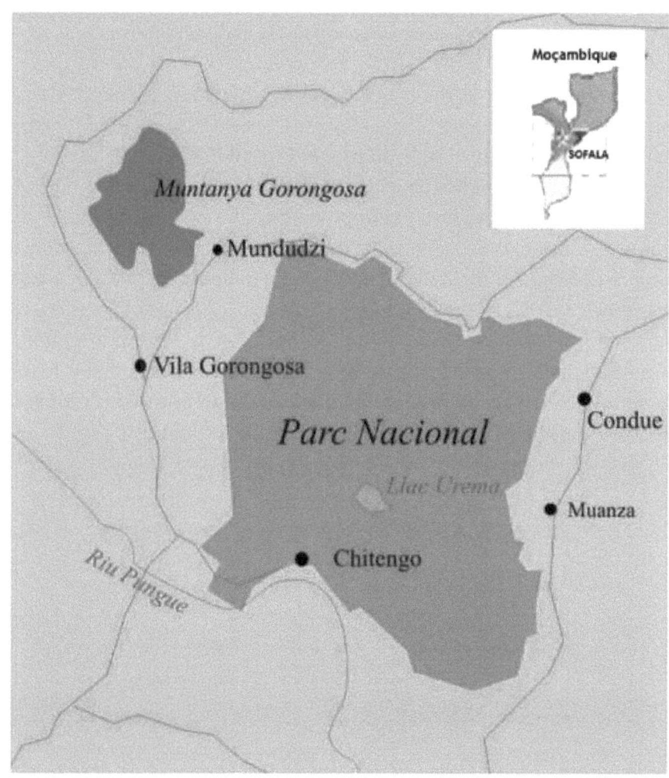

Figure 3. 1 - Location of Mateus Catique Community, Gorongosa (Mozambique)
(Adapted by the author, 2020).

The following figure shows the percentage distribution of dwellings according to the degree of access to basic services.

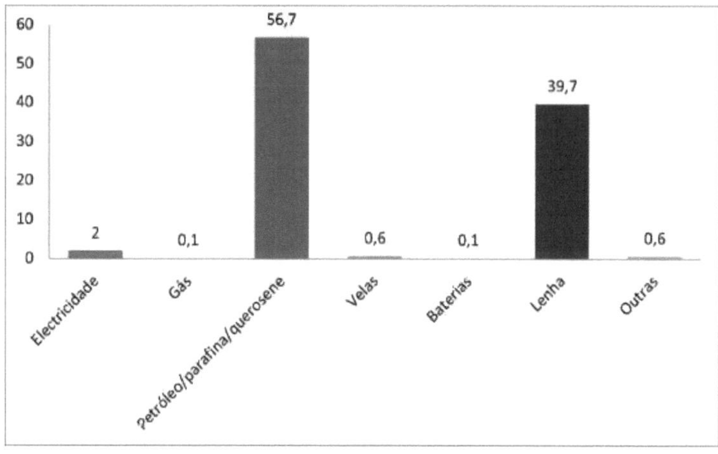

Figure 3.2. - The current form of energy available *(ADG, 2017)*.

This shows that, at district level, the majority of the population lives according to the same rules:

- The main source of energy used by households is oil (57%);
- Around 19 per cent of families have access to drinking water; and
- Around four per cent of families use improved sanitation systems.

The figure above shows that 57% use oil as a source of energy, which is a non-renewable resource and is not environmentally friendly, as it pollutes the air and contributes to greenhouse gases,

And according to the mapping of the district in question, as shown in graph 3.3. which shows the potential that exists in Gorongosa, this work goes into the feasibility and comparative study of 4 types of predominant energy sources in the Gorongosa district, of which hydro, solar photovoltaic, wind and biomass stand out.

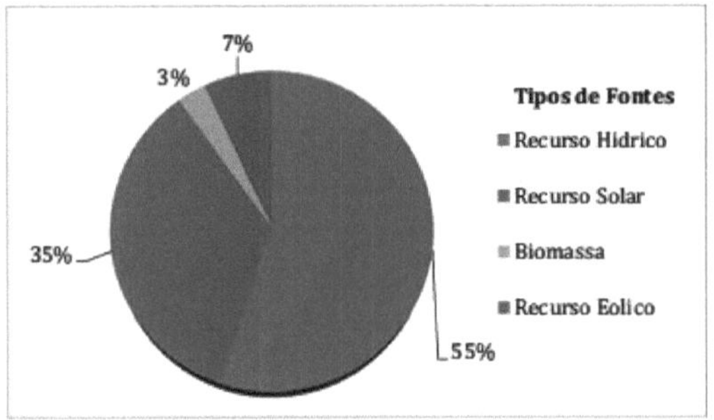

Font types
Water resource
Solar resource
Biomass
Wind resource

Graph 3. 3- Gorongosa's existing potential
(*adapted from:* Namburete, 2014)

3.4. System sizing for renewable energy sources

There are a multitude of types of electrical installations according to the NP, NF and NBR. In this work, the NP will be taken into account for the sizing of two systems with the highest classification in the Gorongosa district, as illustrated above in graph 3.3. the most abundant potential in the district, being hydro and solar.

Therefore, this dissertation will use the installation in accordance with **Resolution No. 10/2009 on new and renewable energies,** which will be selected according to the forecast illustrated in Table 4.2. in Chapter IV, and according to the labour needs of the population of the Vanduzi region, in the Mateus Catique Community in Gorongosa.

- **Survey of current consumption in Gorongosa**

According to ADG (2017), the incidence of ownership of durable goods by families living in the district is shown in the following figure:

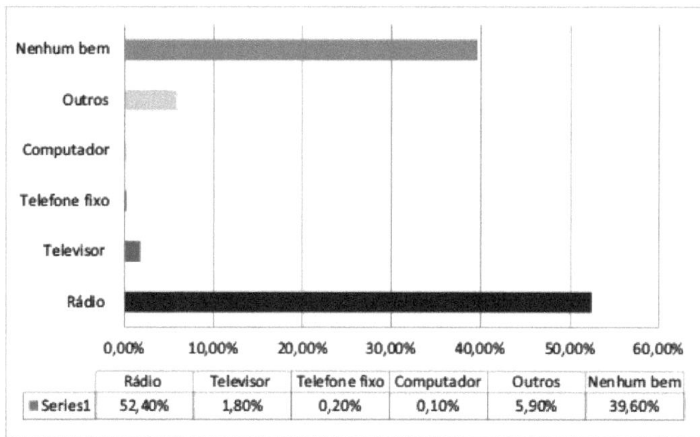

Figure 3.3 - Households, according to ownership of durable equipment
Source: (ADG, 2017)

■ **Energy consumption estimates**

According to Pinho and Galdino (2014), the first task is to identify the number, power and operating time of the equipment that needs to be fuelled.

Annex 6 (table A5) sets out the methodology for the procedure that must be carried out in order to survey daily electricity consumption needs. Thus, the estimated energy is 1196 kW/day.

3.4.1. Hydroelectric system

It should be emphasised that the first step in calculating or sizing a hydroelectric system is to carry out a study of the site where the system is to be inserted, in order to meet the demand of the Mateus Catique community. To do this, the flow rate of the river under study will be determined, taking into account the height of the fall, the area of the site and the speed of the flow, and the author has chosen to take the values by trial and error methods in order to obtain a significant and considerably attractive sample for sizing.

3.4.1.1. Hydrographic data of Gorongosa, Mozambique

According to ADG, (2017) the Gorongosa waterfall, from an economic and social point of view, is a place that can contribute to the district's development on a large scale, insofar as ecotourism, small-scale fishing and the sustainable exploitation of various resources can be developed there.

The waterfall under study is the Nhandare River, which rises in the Gorongosa Mountains, approximately 30 kilometres from the town of Gorongosa. It lies outside the reserve and well outside the territory of PNG. As such, the Nhandare waterfall is quite extensive and approximately 16 metres high at its falls.

3.4.1.2. Survey of hydroelectric potential

According to White, (2001) and Samways, (2004), in order to determine the potential, one must take into account o the concept of flow como being the amount of fluid per unit of time that can pass through the straight section of a pipe or channel.

Thus, according to the height of the drop in the river, the fluid velocity was extracted as a function of time, through tests and observations made by the author, as shown in table

3.2, where 10 observations were made, in order to obtain o the smallest possible error.

Table 3. 2 - Observation for speed calculation

Test	Height (m)	Time (s)
1	16.0	1.90
2	16.0	2.10
3	16.0	3.00
4	16.0	1.90
5	16.0	2.30
6	16.0	1.70
7	16.0	2.20
8	16.0	2.10
9	16.0	1.80
10	16.0	2.05

Source: Author (2020).

From equation 3.1, we determine o average time in order to obtain o average value among the variations found. Where: Tm - average time (s); n - total number of samples.

$$Tm = \frac{\sum_{i=1}^{n} t_i}{n} \quad 3.1$$

Substituting the values into equation 3.1 gives us the following result:
Tm = 2,105s
Thus, at sixteen (16) metres by two (2) seconds, the depth of the river (P) is 3.85 m and the width of the river (C) is 6.45 m, at which point the flow rate (Q) will be determined.

$$Q=h*t*A \quad 3.2$$

Substituting the values into equation 3.2 gives us the following result:
Q = 795 m^3 /s
Thus, 795.00 m^3 /s is the amount of fluid per unit of time that can pass through a turbine without changing the relative height of the initial point of fall.

Figure 3. 4- Waterfall under study, Gorongosa (Author, 2020)

3.4.1.3. Turbine selection

A turbine is chosen to meet certain head values, which depend on the specific conditions in which it is to be installed. This choice also depends on another quantity, which is the nominal rotational speed (rpm) of the electric generator that the turbine will drive.

Determining the specific speed

According to Munson (2002), the parameters for choosing the type of turbine for a given installation are the discharge (Q), the drop (H_{Top}) and the nominal rotational speed (n). With these elements, and o arbitrary preliminary estimate of the efficiency values and existing losses in the system $(\eta_T \text{ e } \eta_H)$. o value of the mechanical power (equation - 3.4) of the turbine ($P_{,nec}$) is calculated and then the choice of turbine is made using equation - 3.3 below.

$$n_s = \frac{n\sqrt{P_{mec}}}{H_{Top}\sqrt[4]{H_{Top}}} \qquad 3.3$$

Substituting the values into equation 3.3 gives us the following result:

n_s = 575 rpm

Thus, from equation 3.4, the mechanical power is determined for the possible choice of turbine.

$$P_{mec} = g^{*}\eta_T{}^{*}Q^{*}H_{Top} \qquad 3.4$$

Substituting the values into equation 3.4 gives us the following result:

P_{mec} - 90.90 rpm

With this value, it is possible to know the power of the turbine shaft of the project, which will be coupled with o generator. Therefore, with this data, the specific speed and the mechanical power on the turbine shaft, the nominal speed can be determined. By substituting the values into equation 3.3, the nominal speed will be: n = 78 rpm

Therefore, looking at the operation of helical or Kaplan turbines according to the NP, it specifies that it operates at a speed corresponding to the range of 50 to 80 rpm. So a Kaplan turbine (Annex 5) with n – 80 rpm, standardised, is chosen.

3.4.1.4. Generator selection

The nominal power of an electric generator (S_n) is mostly the power delivered to the terminals of a given generator. This power is therefore specified in terms of the apparent electrical power according to equation 3.5. However, the choice of generator is based on the dimensioning of rotational speed, rated power both сото the electric power looking at the characteristics, noise limits, frequency, thus сото degree of protection of the generator well adjusted.

= 3_5

Substituting the values into equation 3.5 gives us the following result:

S_n = 140 kVA

For a generator, the base operating frequency is 60 Hz, so an element with even poles must be chosen, as shown in table 3.3, which selects the pole number according to the generator frequency. Disregarding slip, the power required by the load at the base operating point will be:

Table 3. 3 - Selecting the number of poles according to frequency.

60 Hz frequency		50 Hz frequency	
No. of centres	Synchronous rotation	No. of centres	Synchronous rotation

2 Poles	3600 rpm	2 Poles	3 000 rpm
4 Poles	1800 rpm	4 Poles	1500 rpm
6 Poles	**1200 rpm**	6 Poles	1000 rpm
8 Poles	900 rpm	8 Poles	750 rpm

Source: ANEEL, (2003).

One of the ways to reduce the cost of power converter equipment is to build it in a few models, and in accordance with NP. Therefore, since this is a mini hydroelectric plant with low power, we chose the 1306A-E87TAG3 model generator, see Annex 6 (Table A6), with a reduced rpm in order to minimise the cost of the equipment. A generator with a nominal speed of 1200 rpm and 6 poles was chosen.

3.4.1.5. Estimating the cost of fixing the hydroelectric system

Estimating the costs associated with installing hydroelectric systems depends, among other factors, on the power installed, the height of the drop and the connection to the electricity grid.

According to Joao (2016 p. 28), there are data from scholars that allow us to place the total investment in a range between €1,300/kW and €3,750/kW, with the lower limit corresponding to low, medium and high drops and powers above 1,000 kW and the upper limit corresponding to low drops and powers below 500 kW, as can be seen in table 3.1.

Table 3.1 - Estimated costs associated with installing hydroelectric systems systems.

Powers	Minimum (€/kWJ	Maximum (€/kWJ	Average (€/kWJ
1 MW-10MW	600	2000	1300
500 kW-lMW	1300	4500	2900
100kW-500kW	1500	6000	3750
<100kW	1500	6000	3750

Source: Joao (2016). Exchange rate (2020) 1€ = 89.90 MT **Consulted Millennium BIM on 15/11/2020.**

The estimated investment for the construction of the fixed system will therefore be as shown in Table 3.2 below:

Table 3.2 - Estimated fixed investment (system construction) system)

Designation	Cost (MT/kWJ	Demand Power	Investment of the fastening (MT)
Mini hydro plant	337 125.00	210 kW	70 796 250.00

Source: Author (2020).

In order to build this hydroelectric system, the following techniques illustrated in the table must be taken into account, as they take into account the power demand calculated in order to meet all the fixed levels of the elements that make up the system being analysed.

Thus, o dimensioned work is budgeted according to table 3.4 below:

Table 3. 4 - Budget for hydroelectric system equipment, Case of the Mateus Catique Community.

№	Designation	Quantity	Unit Price (MT)	Value (MT)
1	Posts	10	35 000.00	350 000.00
2	+ 3 Fuse 14x51 mm, 690V/4A, class gG (DF)	10	7 000.00	70 000.00
3	200 kVA 6-pole generator	i	800 000.00	800 000.00
4	250 kVA Cu/Al insulated transformer	i	1 825 000.00	1 825 000.00
5	Hydraulic turbine 250 kW	i	1 390 400.00	1 390 400.00
6	ABC CABLES 4x16 mm2	10 000m	60.00	600 000.00
7	PVC CABLES 4mm^2	10 000m	75.00	750 000.00
8	MEDIUM VOLTAGE CABLES	1000m	250.00	250 000.00
9	HIGH VOLTAGE CABLES	1000m	275.00	275 000.00
10	low voltage cables	1000m	260.00	260 000.00
11	Fusible	100	1500.00	150 000.00
12	System construction			70 796 250.00
	Total			**80 625 000.00**
	TOTAL(USD)			**1106 730 USD**

Source: Alibaba.com Exchange Rate (2020) 1USD = 72.85 MT **Consulted Millennium BIM on 15/11/2020.**

3.4.2. Photovoltaic solar system

It should be noted that the calculation of electricity consumption for the Mateus Catique community took into account a population sample of over 50 families, which is approximately more than 5 blocks of the area. Thus, the consumption of electrical equipment and appliances is analysed, in which case we will take into account the power of each piece of equipment, coro can be seen in Annex 2.

3.4.2.1. Climatological data from Gorongosa, Mozambique

According to the ADG, (2017) the municipality of Gorongosa has a predominantly dry tropical climate, with two climatic seasons: the hot and rainy season from November to March, and the dry and cool season from April to October. With an average annual temperature of 22.9º C, with an average annual amplitude of approximately 8 °C (26.°C in November and 18.°C in July), coro shown in table 3.5. provided by the METEONORM 7.2 tool.

Table 3. 5 - Climatic data for Gorongosa district, Mozambique

Location Gorongosa (Mozambique)
Data sourcejMeteoNorm 7.2 station

Irradiation global horizontal kWh/mtmes		Irradiaga diffuse horizontal kWh/rtf.mes	3 Temperature -C	Speed wind m/s	Linke Turbidity H	Relative Humidity
January202	.9	77.7	28.2	3.50	3.200	69.3
February174	.3	75.4	28.2	2.90	3.200	78.7
Margo172 .4		65.6	24.5	3.80	3.600	72.0
April162 .4		51.6	25.7	2.80	2.900	79.9
May144 .8		42.7	23.8	2.50	2.800	78.2
June126 .8		33.9	21.8	3.00	2.700	80.3
July135 .7		42.3	21.7	3.00	2.900	78.1
August157.9		48.0	23.3	3.30	3.200	73.4
September189	.3	74.4	27.7	2.90	3.100	79.7
October198	.3	72.7	26.9	410	3.400	70.3
November 207.5		80.9	29.5	3.00	3.100	79.6
December196	.4	81.0	20.0	3.00	3.200	72.8

Year ?	2068.7	746.2	25.8	3.2	3.108	76.0

Source: MeteoNorm (2019).

3.4.2.2. Solar radiation assessment

Solar radiation in interstellar space is 1353 W/m^2 , known as the solar constant. The energy that reaches the ground on Earth is lower because of absorption in the atmosphere. The amount of solar radiation per unit area that reaches a specific point on Earth depends on the latitude, declination and season (Pinho & Galdino, 2014).

To carry out this solar incidence calculation, climatological data from the municipality of Gorongosa, Mozambique сото shown in table 3.5 above is used, as the inclination of the photovoltaic panels can be adjusted, with the aim of gaining energy production throughout the year, optimising for winter due to the high level of tropical cooling. With this data, it is possible to size the photovoltaic system.

Table 3. 6 - Total monthly and annual solar incidence

Mes	Average (radiant hours)	Days	Monthly solar incidence (ir r)
January	6.55	31	202.9
February	6.23	28	174.3
Marfo	5.56	31	172.4
April	5.41	30	162.4
May	4.67	31	144.8
June	4.23	30	126.8
July	4.38	31	135.7
August	5.09	31	157.9
September	6.31	30	189.3
October	6.40	31	198.3
November	6.92	30	207.5
December	6.34	31	196.4
Total (Annual)			**2068.7**

Source: MeteoNorm (2019).

According to data provided by METEONORM 7.2, the average global horizontal irradiation for the Gorongosa district corresponds to 2068.7 kWh/m^2 /year. As table 3.6 shows, it can also be seen that the most critical months, i.e. those with the lowest irradiation values, correspond to the winter months.

3.4.2.3. Determining the optimum solar incidence

The inclination of the photovoltaic panels directly affects the energy produced by the photovoltaic system.

0 Gorongosa district extends between longitudes 33° 45' E and 34° 40' E and latitudes 18° 15' S and 19° 00' S, and is located in the centre-western part of Sofala Province (ADG, 2017).

Latitude is a datum that we extracted from ADG, (2017), referring to the Climatological Data of the district in question. This latitude for the Gorongosa district is **delS.15**°.

The inclination of the system in the winter period is given by equation - 3.6 below.

$$\theta_{inv} = latitude + 15° \quad (3.6)$$

Substituting o value into equation - 3.6 we get the following result:

$\theta_{inv} = 33{,}15^{\circ}$

The optimum solar inclination of the system in the summer period is given by equation - 3.7 below.

$$\theta_v = latitide - 15^{\circ} \qquad 3.7$$

Substituting o value into equation - 3.7 we get the following result:

$\theta_v = 3{,}15^{\circ}$

Thus, to determine the optimum solar incidence, the heating season (summer) is evaluated, because this is the period with the highest profitability for solar panels and because of the high level of heating in the area.

Since our source depends on this climatic factor, we need to determine this optimum solar incidence, which is represented by [f(0p)] with equation - 3.8 below:

$$I(\theta_v) = \frac{Incid\hat{e}ncia\ Solar\ Tabelada\ (IST)}{[1-(4.46+1.19)*(0.0001*\theta_p)]} \qquad 3.8$$

Substituting the values into equation - 3.8 we get the results in table 3.7:

Table 3. 7 - Total monthly and annual optimum solar incidence

Mes	Average (hours of radiation)	Monthly solar incidence (It)	Optimum monthly solar incidence (i^r r)
January	6.55	202.9	204
February	6.23	174.3	175
Marfo	5.56	172.4	173
April	5.41	162.4	163
May	4.67	144.8	147
June	4.23	126.8	129
July	4.38	135.7	136
August	5.09	157.9	159
September	6.31	189.3	190
October	6.40	198.3	199
November	6.92	207.5	208
December	6.34	196.4	197
Annual Total		**2068.7**	**2079**

Source: Author (2020).

3.4.2.4. Determination of demand power for Mateus Catique community, Gorongosa

0 monthly electricity consumption of the Mateus Catique community, Gorongosa, is up to 602 kWh, caused by the operation of equipment at full load provided for residents, as well as the provision of additional load for them.

By cross-referencing the information in Annex 2, the Average Monthly Energy Consumption, and taking into account a population sample of more than 50 families, which is approximately more than 5 quarters of the region, consumption rises to 34939.20 kWh.

It is then possible to determine the number of solar panels needed for the annual supply, in order to guarantee the proper functioning of the community's equipment.

Total annual energy (WTP) = 34939.2 kWh/month x 12 months

WTP = 419270.40 kWh/year

The power required is the ratio of the plant's total annual energy (energy to run the

equipment in the Mateus Catique community in Gorongosa) to the total annual solar incidence.
So, once we know the total annual energy available to generate electricity in the community, we determine the power demand of the site according to equate - 3.9 shown below.

Power demand (PD) $= {ETA}/{I(\theta_v)}$ 3.9

Substituting o value into equation - 3.9 we get the following result:
PD = 202 kW
The system will then be sized for this power value, which will meet the demand required due to the rational use of the equipment in each household in the Mateus Catique community in Gorongosa.

3.4.2.5. Selection of solar panels

According to Solax *Power (2020),* South Africa's solar panel manufacturer (solar panel dealers), the solar panels the company owns have a power output of up to 300W.
Therefore, for this study, the same solar panels supplied by BUNDU POWER SOLAX were taken into account. This company was chosen because it is a neighbour of the country (Mozambique) and because of the quality of the solar panels supplied by the company, with a maximum electrical power (PSV) of 300 W, as shown in table 3.8.
Thus, the quantity of solar panels needed to meet the average annual load, taking into account the month of the year ratio and meeting the requirements for surplus energy, is given by the following equation - 3.10:
Panel output power for sale - PSV

Number of solar panels (QPS) $= {PD}/{PSV}$ 3.10

Substituting o value into equation - 3.10 we get the following result:
QPS = 672 solar panels.
Bearing in mind the losses that can occur within electrical installations, whenever necessary, the author opted for the method of adding values, and so the number used for the system is 680 solar panels grouped together (40x17) as shown in **Annex 3.1,** as a way of reducing the losses that can occur within our system.

Table 3. 8 - Electrical characteristics of the panel, model BP300WP

Modulo Electrical Parameters	
Maximum Electrical Power	300W
Maximum Power Current	8,33A
Maximum Power Voltage^mx]	36,0 V
Short-circuit current	9,0A
Open circuit voltage	45,0 V
Maximum system voltage	1000 V
Weight	25.0 kg
Dimensions	1946 x 980 x 40 mm
Voltage Temperature Coefficient	- 0.37 V/K

Source: Solax Power (2020).

3.4.2.6. Inverter selection

When choosing inverters, account must be taken of the energy regulation referred to in this chapter, and taking into account the standardisation of the reference temperature of photovoltaic modules, it is assumed according to NP and NF: the Reference temperature of the district, which is 25⁰ C = 298.15 KeThe solar panel has a voltage temperature coefficient of $-0{,}37\frac{V}{K}$.. From equation - 3.11, the adjustable voltage of the photovoltaic modules is determined.

$$V_{max} = 45{,}0 * \{(1 + [T_{min} - T_{ref}] * (-0{,}37\tfrac{V}{K}))\} \qquad 3.11$$

From Table 3.4, according to the MeteoNorm 7.2 software, we take the minimum temperature, which is 21.7 °C = 294.85 Л", and taking into account the referential or ambient temperature, we calculate the adjustable voltage, and we obtain the following result:

$$V_{max} = 106\ V$$

Therefore, this work proposes the use of the inverter shown in figure 3.6 below, model 1260WD3-HV620 *GPTech Inverter*, which is protected by a galvanic layer and has active temperature control technology.

To determine the maximum number of inverters, equation 3.12 is taken into account, and to know the exact number of modules to be installed, table 3.9 shows the electrical characteristics of the inverter to determine the maximum number of inverters.

Table 3. 9 - Electrical characteristics of the inverter model 1260WD3-HV620

Electrical characteristics	
Maximum DC power (cos ϕ=1)	15000 kW
Maximum input voltage	1500 V
MPP - voltage range	895 V- 1250 V
Maximum current	10000 A
OUTPUT (AC)	
Nominal power	830 kW
Output voltage (range)	180 V-620 V
Frequency range	50Hz-60Hz
Maximum current	1334,0 A
EFFICIENCY	
Maximum efficiency	98,80 %
GENERAL DATA	
Dimensions (W x H x D)	390x495x127 mm
Weight	23.5kg
Operating temperature - 20 °C to +60 °C or - 4°F to 140°F	
Altitude	<1000m

Source: Solax Power & GPTech (2020).

$$Q_{max} = \frac{U_{dc}}{V_{max}} \qquad 3.12$$

Substituting o value into equation - 3.12 we get the following result:

Qmax = 5 modules

To determine this, the minimum city temperature for the project is taken into account, and this becomes the operating temperature to help determine the number of inverter modules to be used in the system. For a maximum voltage of 106 V per module, a maximum of 5 modules can be connected.

Thus, o dimensioned work is budgeted according to table 3.10 below:

Table 3. 10 - Equipment budget for the photovoltaic system, Case of the Mateus Catique Community.

Nº	Designation	Quantity	Unit Price (MT)	Value (MT)
1	Portafusible socket 14x51 mm, 3P/690V/50A (EATON)	10	5 000.00	50 000.00
2	+ 3 Fuse 14x51 mm, 690V/4A, class gG (DF)	10	7 000.00	70 000.00
3	300W solar panels	680	44 000.00	29 920 000.00
4	GPTech DC/AC inverters	3	117 000.00	351 000.00
5	ABC CABLES 4x16 mm2	10 000m	60.00	600 000.00
6	PVC CABLES 4mm²	10 000m	75.00	750 000.00
7	32 A circuit breaker	200	350.00	7000.00
8	System construction			20 000 000.00
	Total			**55 970 000.00**
	TOTAL(USD)			**770,000 USD**

Source: Solax Bundu Power Store, Exchange rate (2020) 1USD = 72.85 MT **Consulted Millennium BIM on 15/11/2020.**

In this chapter we have presented a global survey of data, considered the various important factors that influence the design of the new matrices planned for the Mateus Catique community in Gorongosa.

Starting with the hydroelectric system, which occupies around 35% of the sources in the Gorongosa district, analysing: the type and capacity of the equipment, the conditions and quantity of the flow, among others, with a particular eye on the degree of resistance, in order to properly select the equipment to be used in the system.

Then the solar photovoltaic system, analysing: working conditions, equipment consumption, climatological data, among others, with a particular view to its annual or monthly balance, for the appropriate selection of photovoltaic modules to be used in the system.

3.5. Energy Demand Model

Regression analysis for energy forecasting and demand over the years, using the simple and most widely used equation, is given by equation 3.13:

$$LnEt = a + bLn(Pe) + cLn(Yt) + ut; \quad 3.13$$

3.5.1. Description of Model Variables

The coefficients estimated here are price and production elasticities of energy demand, which measure the percentage changes in industrial energy demand, residential per percentage change in energy prices and economic output.

And_t = energy demand;

Yt = income;

P_e = energy price;

u_t = error.

3.5.2. Importance of the Model

This model is important because it is an econometric model and has the following characteristics:

- Used to study a complete and homogeneous class of consumers.

- They don't take into account their technological structure and the end-use of energy.
- Represent o consumption by an equation.
- They require less data.

4. Results obtained

4.1. General

The Mateus Catique community is made up of villages (houses) that are relatively close to each other and where the houses are organised into plots distributed along a main road, where the residents are fundamentally dedicated to subsistence farming and some of them carry out their activities in the PNG. Unlike the neighbouring communities como: Nhamacola and Casa Banana, in this community the buildings have different spatial organisation characteristics.

The inhabitants of these communities, Nhamacola and Casa Banana, are fundamentally dedicated to subsistence farming and the villages are very dispersed. And in terms of allocating electricity supply projects, there will be a great deal of expense in buying electricity cables. Due to the way the villages are organised along an unparceled road.

Table 4.1 - Summary of Results

Principals Parameters	Hydropower	Solar Energy	Remarks
Drop height	16 m	-	Low fall
Demand power	202 kW	202 kW	Mini centre
Generator	253 kVA/202.3 kW	-	Perkins - 1306A E87TAG3
Transformer	250 kVA	-	Cu / Al insulated
Turbine	250 kVA	-	Hydraulics
Summer slope	-	3.15°	Warm-up period
Number of panels	-	672 300W panels	BP300WP, Solax Power
Adjustable voltage	-	106 V	5 Inverters, 1260WD3 - HV620
PVC cables	4 mm²	4 mm²	System connection

Source: Author, 2020.

4.2. Analysing social viability

4.2.1. Social characterisation of the solar photovoltaic system and the hydroelectric system

Forjaz (2011) and Sebastiao (2016) define rural electrification as a set of actions created to provide and supply electricity to inhabitants of areas with specific characteristics, such as the Mateus Catique Community, including small energy loads and the creation of specific opportunities in these areas.

The system for supplying electricity to this area varies and can include isolated generators that serve collective or individual consumers (Annex 4 - Connection to customer switchboard). However, the variation in these methods depends on local circumstances and the degree of saturation of the electricity supply.

The electrification of rural areas in developing countries is complex and requires a great deal of specific skills and equipment, as shown in chapter 3 of this dissertation. The impartial planning, realisation and operation of rural electrification projects cannot be separated from dilemmas such as poverty, concerns about environmental degradation, rural development and energy needs in general.

It should be emphasised that although there are limitations of a social nature, the application of these tools brings social and economic advantages to the inhabitants of these areas.

One of the main objectives of rural electrification is the pursuit of economic development with a view to increasing agricultural production in the community in question, as well as at the level of the district and the central region of Mozambique, and it should be assessed differently for reasons of a purely social nature.

The fundamental impact of rural electrification is that it transcends local physical boundaries, providing social and environmental benefits, as will be shown below: it changes local habits, improves people's living conditions, attracts industry to the region and boosts local trade, as can be seen in figure 4.1 on the following page.

The figure below illustrates the interaction between social analysis, economics, public services and rural modernisation (the main objective of rural electrification), which is the main instrument for local, regional and national socio-economic development.

Thus, the social analysis that these two systems, already designed to meet the population demand of the Mateus Catique Community, are aimed at improving the living conditions of the local population, including the creation of suitable conditions for education, agriculture, fishing and health, are fundamental elements for the development of a given area.

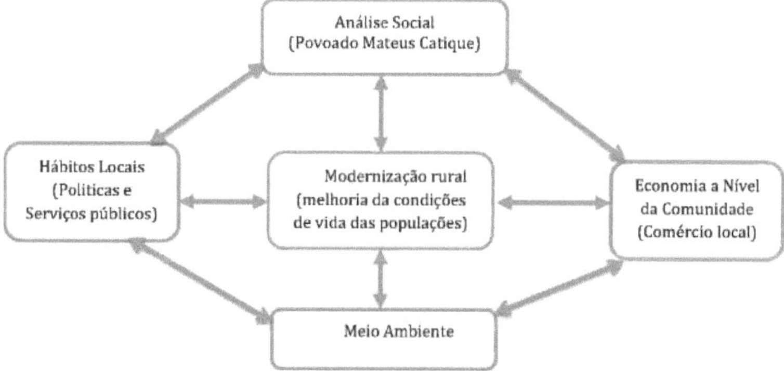

Figure 4.1 - Impact of electrification in rural areas, Case of the Mateus Catique Community.

Source: Freitas et al., (2020 p. 20), Apud Sebastiao, (2016).

In other cases, rural electrification is a factor for society, it must go hand in hand with, rather than accompany, regional development. In both cases it is planned and implemented in combination with other sustainable rural development activities, under certain social and political circumstances, and its success also depends on the success of other activities.

With the integration of these systems in the Mateus Catique community, socially they bring with them various advantages, as demonstrated above, the fundamental element of which is rural electrification, and this being one of the key factors that is part of the five-year plan of the Government of Mozambique, and this research contributes to this plan, bringing with it a new model and characteristics of two systems that the district of Gorongosa currently offers for the first phase: lighting and acceptance in the area of its use in the agricultural and livestock sector.

4.3. Analysing the economic impact of the investment

Solar-photovoltaic systems depend for the most part on the quality of the sun in a given city, and the district of Gorongosa has good sunshine throughout the year. After sizing the systems according to the justified energy matrices, the following results are obtained (see table 4.1).

Table 4.1 - Main results of the dimensioned system

System	Demand power	Investment (MT)	Classification
Hydroelectric	210 kW	**80 625 000.00**	Mini hydroelectric power station
Solar - Photovoltaic	210 kW	**55 970 000.00**	Mini solar power station

Source: Author, 2020.

O district has a good capacity for installing hydroelectric systems, as most of the rivers are part of the PNG area, and their studies depend on the direction of the park. Not only because of this factor, it is known that hydroelectric systems, despite being a renewable source and not emitting pollutants, cause major environmental and social impacts, cause the extinction of certain marine specialities and make the environment conducive to the transmission of diseases such as malaria and others.

For these reasons, it would be difficult to study all the rivers in the district, which represents 35% of the water resource. Therefore, the author opted for a solar photovoltaic system for the site under study, as it does not cause fraudulent environmental problems compared to hydropower and the district has a good capacity for utilising solar photovoltaic systems.

A photovoltaic system is recommended because, in terms of technology, solar systems don't require a lot of manpower to maintain over time, and the mini hydroelectric plant dimensioned in this work, there was no experimental and laboratory analysis of the Potential Hydrogen (PH) level of the Nhandare River water in order to determine the lifetime of the turbine blades and the cost analysis for implementing the mini hydroelectric power station, which is why the author suggested installing the solar photovoltaic system for the Mateus Catique Community.

4.3.1. Economic characterisation of solar photovoltaic and hydroelectric systems

To meet a demand of approximately 60 families, which corresponds to a power of up to 210 kW, o which corresponds to a total grouping of 40x17 solar panels and each device with an output power of 300 W, the connection being in mixed circuit, i.e. a part with serial and parallel connection, como shown in annex 3 - general connection of the photovoltaic system.

Daily electricity consumption is expected to be between 1.OkWh and 3.5 kWh and each kWh is equivalent to 8.00 MZN (EDM, 2019). Therefore, this work has taken into account the following developed information, as illustrated in tables 4.2 and 4.3 below.

Table 4. 2 - Forecast of residential electricity consumption, Case of the Mateus Catique Community.

Year	Daily Forecast Total (kWh)	Total monthly forecast (kWh)	Total annual consumption forecast (kWh)	Amount collected for consumption (MT)
2020	105	3150	37805	302440

2021	120	3600	43205	345640
2022	141	4230	50765	406120
2023	145.5	4365	52385	419080
2024	175.5	5265	63185	505480
2025	190.5	5715	68585	548680
2026	193.5	5805	69665	557320
2027	196.5	5895	70745	565960
2028	211.5	6345	76145	609160
2029	217.5	6525	78305	626440
2030	226.5	6795	81545	652360

Source: Author, 2020.

Table 4. 3 - Electricity consumption forecast data, Case of the Mateus Catique Community.

Year	Energy to be Billed (kWh)	Population growth	IPC	GDP (Private Consumption) [10^6 MT)
2020	37805	184395	117.85	615853.216
2021	43205	192104	119.37	651503.990
2022	50765	199813	120.90	687152.443
2023	52385	207522	122.42	722801.669
2024	63185	215231	123.94	758450.896
2025	68585	222939	125.47	794100.122
2026	69665	230649	126.99	829749.348
2027	70745	238358	128.52	865398.575
2028	76145	246068	130.04	901047.801
2029	78305	253777	131.56	936697.028
2030	81545	261487	133.09	972346.254

Source: Author, 2020.

The variables CPI (Consumer Price Index) and GDP (Gross Domestic Product).

The CPI variable is aimed at carefully reading the consumption generated by the equipment over a given period of time, and this variable has the function of assessing the purchasing power of the consumer, i.e. the residents over the years, see table 4.2 above.

Table 4. 4 - CPI forecast as a function of time for 2020 to 2030.

SUMMARY OF RESULTS								
Regression statistics								
Multiple R	0.516792							
Square of R	0.267074							
Adjusted R-squared	0.120489							
Squad error	5.974102							
Remarks	7							
ANOVA								
	Gl	*SQ*	*MQ*	*F*	*F for meaning*			
Regression	1	65.02603	65.02603	1.821973	0.234967			
Residual	5	178.4495	35.68989					
Total	6	243.4755						
	Coefficients	*Error-standard*	*Stat t*	*P-value*	*95% lower*	*95% Bottom*	*top 95.0%*	*Superior 95.0%*
Intercept	-2960.49	2274.934	-1.30135	0.249873	-8808.39	2887.417	-8808.39	2887.417
Variable X 1	1.523929	1.128999	1.349805	0.234967	-1.37826	4.426113	-1.37826	4.426113

Source: Author, 2020.

Once the CPI values of previous years are known according to INE data, it is possible to predict future years, as illustrated in table 4.4 above. The known data refers to the years 2012 to 2018 (totalling 7 known observations). Thus, using the regression technique, it was possible to determine the future years.

Table 4. 5 - GDP forecast as a function of time for 2025 to 2035.

Regression statistics								
Multiple R	0.976010252							
Square of R	0.952596012							
Square of R adjusted	0.945824014							
Squad error	23282.5087							
Remarks	9							
ANOVA								
	Gl	*SQ*	*MQ*	*F*	*F for signifiedneia*			
Regression	1	7.63E+10	7.63E+10	140.6669	6.88E-06			
Residual	7	3.79E+09	5.42E+08					
Total	8	8E+10						
	Coefficients	*Error-standard*	*Stat t*	*P-value*	*95% lower*	*95% higher*	*Lower 95.0%*	*Superior 95.0%*
Intercept	-71395583.34	6050598	-11.7998	7.12E-06	-8.6E+07	-5.7E+07	-8.6E+07	-5.7E+07
Variable X 1	35649.2264	3005.759	11.86031	6.88E-06	28541.74	42756.72	28541.74	42756.72

Source: Author, 2020.

The GDP variable aims to provide an internal reading of the different types of products and services that the producers in the "Mateus Catique Community" district produce each year, and this aims to evaluate the sum of all goods and services in the region. The known data refer to the years 2009 to 2017 (totalling 9 known observations). Thus, using the regression technique, it was possible to determine the future years as shown above in table 4.5.

Table 4. 6 - Complete electricity consumption forecast data, Case of the Mateus Catique Community.

QCPW K

Year	Energy to be billed (kWh)	Population growth	IPC	GDP (Private Consumption) [10^ 6 MT1	T	ln(Q)	ln(CP)	ln(W)	ln(K)
2020	37805	184395	117.85	615853.216	1	10.5402	12.12484	4.769413	13.33076
2021	43205	192104	119.37	651503.990	2	10.67371	12.16579	4.782228	13.38704
2022	50765	199813	120.90	687152.443	3	10.83496	12.20514	4.794964	13.44031
2023	52385	207522	122.42	722801.669	4	10.86638	12.24299	4.807458	13.49089
2024	63185	215231	123.94	758450.896	5	11.05382	12.27947	4.819798	13.53903
2025	68585	222939	125.47	794100.122	6	11.13583	12.31465	4.832067	13.58496
2026	69665	230649	126.99	829749.348	7	11.15145	12.34865	4.844108	13.62888
2027	70745	238358	128.52	865398.575	8	11.16684	12.38153	4.856085	13.67095
2028	76145	246068	130.04	901047.801	9	11.24039	12.41336	4.867842	13.71131
2029	78305	253777	131.56	936697.028	10	11.26837	12.44421	4.879463	13.75012
2030	81545	261487	125.47	794100.122	11	11.30891	12.47414	4.891026	13.78747

Source: Author, 2020.

CP - Population growth, W - represents o CPI, K - represents o GDP, Q - represents o Consumption. Thus, we determine the regression analysis of consumption as a function of the variables under consideration, in this case o CP, CPI and GDP, in order to understand the behaviour of consumption como illustrated in table 4.6.

Table 4. 7 - Results data for electricity consumption as a function of CP, CPI and

GDP, Case of the Mateus Catique Community.

SUMMARY OF RESULTS								
Regression statistics								
Multiple R	0.993149							
Square of R	0.986345							
Adjusted R-squared	0.980493							
Squad error	0.035747							
Remarks	11							
ANOVA								
	Gl	*SQ*	*MQ*	*F*	*F for meaning*			
Regression	3	0.646113	0.215371	168.5464	6.89E-07			
Residual	7	0.008945	0.001278					
Total	10	0.655057						
	Coefficients	*Squad error*	*Stat t*	*P-value*	*95% lower*	*95% higher*	*Bottom 95.0%*	*Superior 95.0%*
Intercept	46.93736	126.2243	0.371857	0.721001	-251.536	345.4103	-251.536	345.4103
Variable X 1	-32.0828	156.9855	-0.20437	0.843881	-403.295	339.1289	-403.295	339.1289
Variable X 2	2.582319	144.3005	0.017895	0.986222	-338.634	343.7987	-338.634	343.7987
Variable X 3	25.52571	81.76354	0.312189	0.763991	-167.814	218.8658	-167.814	218.8658

Source: Author, 2020.

From equation - 3.13 given in chapter III, the consumption equation is represented as a function of the independent variables population growth, o Consumption price index and gross domestic product.

Thus, o the search model found will be:

$$\mathbf{Ln(Q) = 46.93736 - 32.0828*ln(CP) + 2.582319*ln(W) + 25.52571*ln(K)}$$

Currently, around 41 per cent of the population in the Mateus Catique community use biomass for survival, and 57 per cent use other types of electricity sources such as candles, oil, paraffin, paraffin, gas and others (ADG, 2017), as can be seen in graph 4.1.

Current situation of the Mateus Catique Community (2020)

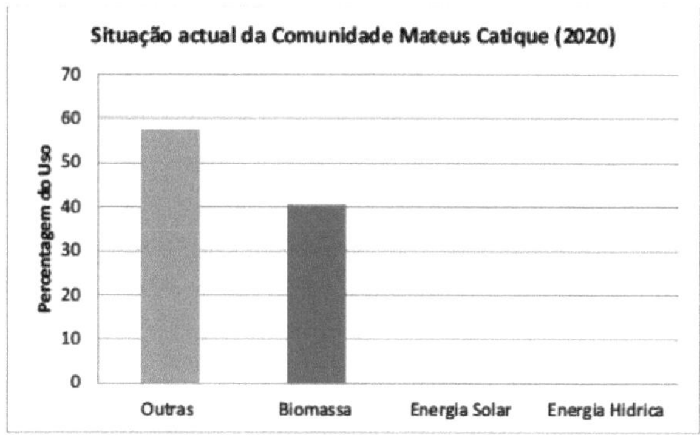

Graph 4.1 - Current situation of energy use in the Mateus Catique Community, Gorongosa (Author, 2020).

Use situation in the Mateus Catique Community (2025)

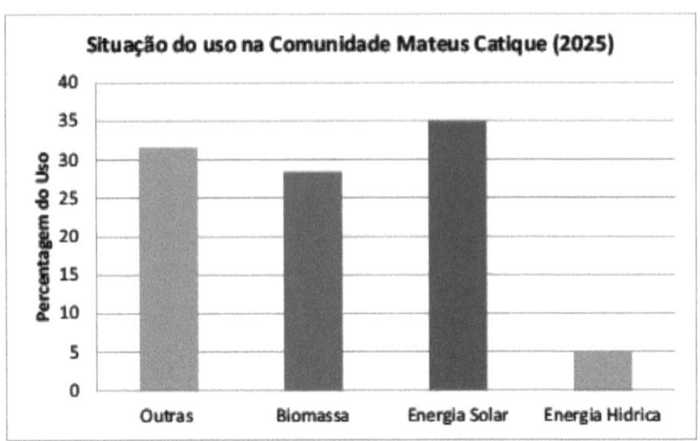

Graph 4.2 - Expected energy use situation 2025 in the Mateus Catique Community, Gorongosa (Author, 2020).

O graph 4.2 shows that with population growth in the region of Gorongosa, and from the ADG, (2017) the Vanduzi administrative post, will be one of the posts that will contribute to the greater demand for energy, which could increase the pressure on biomass in addition to the hydro and solar photovoltaic resources.

Forecast of energy use in the Mateus Community Catique (2030)

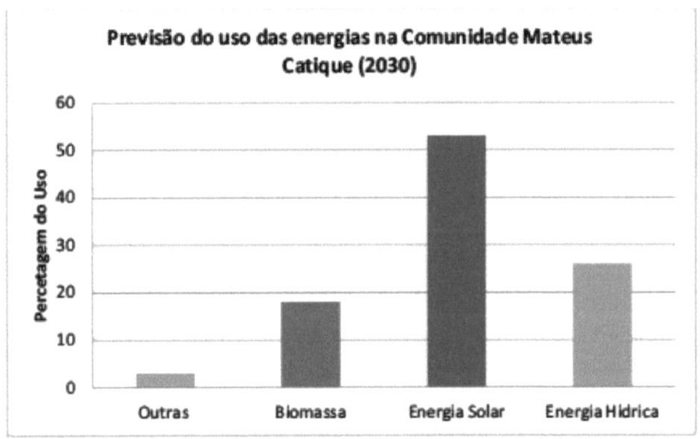

Graph 4.3 - Expected situation of energy use 2030 in the Mateus Catique Community, Gorongosa (Author, 2020).

And by 2030, with the government's project "to fulfil the 2030 agenda - energy for all" and according to this growth, the demand for new local sources in this district is notable, as well as a large increase in photovoltaic solar energy over the years and, this growth, causing a decline from initially (in 2020) 40.5% to 17% for biomass demand (in 2030). The author therefore decided to design an alternative energy system in order to reduce the use of non-renewable energies, since they contribute negatively to the environment.

Analysing the data

46.93736 - Represents the expected amount of energy to be billed, 46.93736 Kilo-Watt-hours, when the independent variables (CP, W and K) are constant.

-32.082% - This will represent the expected rate of decrease in the energy billed (Q) due

to the variable (CP) while keeping the other variables (W and K) constant.
2.582% - This will represent the expected rate of increase in the energy billed (Q) due to the variable (W), keeping the other variables (CP and K) constant.
25.525% - This will represent the expected rate of increase in billed energy (Q) due to the variable (K) while keeping the other variables (W and K) constant.

On the correlation between the variables

R^2 = 0.986345*100% = 98.63% - 0 which means that 98.63% of the variation in billed energy (Q) can be explained by the variation in the variables CP, W and K and 1.37% of the variation in billed energy can be explained by other variables not analysed.
R multiple = 0.993149 - 0 which shows that there is a strong correlation between the dependent variable and the independent variables.

CHAPTER 5

5. Conclusion

5.1 General conclusions

Through bibliographical research on alternative energy sources, it was noted that there are many studies in the area and that the African population has developed systems to guarantee access to electricity for various communities, in order to fulfil the Millennium Development Goals (universal access to lighting) and Mozambique has a high level of implementation of renewable projects compared to some developed countries. However, the aim was also to find out the total power generated by hydroelectric power and other electricity from renewable sources in the Gorongosa District and Catique Community. Therefore, the study was based on defining energy demand, finding suitable technologies and the potential of economic activities at the level of the Catique Community.

Having mapped the Gorongosa district, it can be seen that it has a wide range of renewable resources, the most prominent of which are: biomass energy with around 3%, wind energy with 7%, solar energy with 35% and hydropower with 55%, with approximately 35% of the hydropower resource being found in the PNG reserve area and the rest in areas outside the reserve, such as the Nhandare River.

Mapping at district level showed that water, solar and wind resources are the most abundant in the region. The calculations for sizing the solar photovoltaic system for the Mateus Catique community, for a demand of 202kW. The system consists of 672 300 W solar panels, 5 DC/AC inverter modules, 10,000 metres of ABC 4x16 mm^2 and PVC 4 mm^2 cables, among other connection elements. Unlike the hydroelectric system, with a capacity of up to 140 kVA on the Nhandare River, this is a mini hydroelectric power station because the installed power is less than 0.5 MW and the drop height is a low drop of 16 metres, i.e. in the 2 - 20 metre drop range. The system consists of a 200 kVA generator, a 250 kVA transformer, 10 poles, 1,000 metres of high, medium and low voltage cables, and other elements for connecting the system.

Finally, having analysed the energy matrices taking into account population growth, private consumption and the consumer price index in relation to the photovoltaic and hydroelectric systems already dimensioned, it can be seen that for the next 10 years, there will be a reduction in the use of biomass, and an exponential growth in the use of solar photovoltaic systems, as this system will reduce the level of pollution in the environment, contributing to the sustainable development of the Mateus Catique Community.

5.2 Future work

The development of this dissertation by its nature always leaves some gaps that could result in more in-depth research in certain areas of knowledge. Here we highlight some aspects that could be studied in the future, namely:

- Sizing a solar tracker.
- Feasibility study of the project taking into account the payback period;
- Analysis of local data for the design of a hybrid system - photovoltaic - hydroelectric - taking into account the main environmental risks in the community under study;
- Development of a detailed energy matrix for Sofala province, taking into account the priorities and challenges of Gorongosa district.

Bibliographical references

- GORONGOSA DISTRICT ADMINISTRATION (ADG). Gorongosa District Strategic Development Plan. Reserved articles on the Gorongosa District website, Mozambique for all, 2017.
- ANEEL. Entrepreneur's guide to small hydroelectric power stations. Brasilia, 2003.
- BERNAL, Jonathas Luiz de Oliveira - Sizing a Photovoltaic System. TECHNICAL-SCIENTIFIC REPORT - SAO PAULO - 2014.
- BOLETIM DA REPUBLICA - Resolution no 10/2009 "Approves the Energy Strategy and revokes Resolution no.[2] 24/2000, of 3 October" OFFICIAL PUBLICATION OF THE REPUBLIC OF MOZAMBIQUE, 4 June 2009.
- DA SILVA, Romeu - Mozambique bets on renewable energies - Edited by Nadia Issufo/Renate Krieger, Permanent link https://p.dw.com/p/12GsE, 15.08.2011.
- DE CASTRO, Nivalde Jose; MARTINI, Sidnei; BRANDAO, Roberto - Importance of Alternative and Renewable Sources in the Evolution of the Brazilian Electricity Matrix, V Seminar on Generation and Sustainable Development organised by Fundacion MAPFRE, 2009.
- Electricidade de Mozambique (EDM) - Electricity tariff systems. Public Company (EP), 2019.
- ELLIOT, D. Renewable Energy and Sustainable Futures. Vol 32, pp261- 274. Great Britain, 2000.
- FORJAZ, Jose, et, al. (2011). Sustainable Architecture in Mozambique. Manual of Good Practices. IDG Imagem Digital Grafica, CPLP. Lisbon.
- FORTES, Antonio Gonzalves; MUTENDA, Francisco Mubango; RAIMUNDO, Baltazar - Renewable energies in Mozambique: availability, generation, use and future trends, Revista Brasileira Multidisciplinar, ISSN: 1415-3580, e- ISSN: 2527-2675; DOI:10.25061/2527-2675/ReBraM/2020.v23il.681. 2020, http://revistarebram.com/index.php/revistauniara
- FRANCA, Carlos Renato - Study management, conventional and alternative energy sources, a reflection on energy alternatives, CEMIG-2001.
- FUNAE - Fundo de Energia; Strategic Plan 2010-2014, Maputo, April 2009.
- FUNAE - Energy Fund; Renewable energy project portfolios - hydro and solar resources, 2-. Edifao. Maputo, July 2019.
- INE (2017), Annual projections of the total, urban and rural population of the districts of Sofala province 2007 to 2040. National Statistics System. Mozambique, 2017.
- JOAO, Manuel Raul - Economic feasibility study of mini-hydroelectric and *diesel-electric* power station projects for Majaua, Zambezia province (Mozambique), ISEP, Master's Degree in Electrical Engineering, August 2016.
- LAKATOS, Eva Maria and MARCONI, Marina de Andrade. Scientific Methodology. 4th Revised Edition. Sao Paulo: Atlas, 2007.
- MENDES, Michael; RAMOS, Ana Ferreira (2018). *Construction Characterisation of PNG Villages.* Master's dissertation in Sustainable Construction, UniZambeze - October, 2018.
- MINISTRY OF ENERGY - Strategy for the Conservation and Sustainable Use of Biomass Energy (ECUSEB), Republic of Mozambique - April 26, 2013.

■ MUNHOZ, Guilherme, Study of a turbine for implementation in a hydroelectric power station. Course Final Paper - 2015

■ MUNSON, Bruce R and YOUNG, Donald F "Fundamentals of Fluid Mechanics, 4th ed. America 2002".

■ NAMBURETE, Salvador - Mozambique Renewable Energy Atlas - resources and projects for energy production - 2014.

■ OLIVEIRA, C.E.L.; FRUHLING, I.; URIBE-OPAZO, M.A. Climatological analysis of wind potential in the region of Cascavel - PR. Engenharia Agricola - Jaboticabal, v. 23, n. 3,p.425-433,2003.

■ PATEL, Mukund R. (2006). *Wind and solar power systems: design, analysis, and operation* (PDF) 2- ed. United States of America: Taylor & Francis. Consulted on 17 January 2020.

■ PINHO, Joao Tavares & GALDINO, Marco Antonio - Engineering Manual for Photovoltaic Systems. CEPEL-CRESESB, Revised and Updated Edition - Rio de Janeiro - March 2014.

■ PINTO, I. Elaboration of a Wind Atlas for Mozambique using the RegCM Regional Climate Model. Degree work. Mozambique, 2008.

■ ROCHA, Vinicius Luis - Casa Eficiente, Bioclimatologia e Desempenho Termico - Volume I and II, Florianopolis, UFSC, 2010.

■ SAMWAYS, Carlos Alberto, "Study of the implementation of a small hydroelectric power station in the urban area of ponta grossa/pr", Curitiba 2004.

■ SANTOS SERRAO - Photovoltaic solar energy: case study in the lighting system of a vocational school - PUC, Campinas, 2010.

■ SCHEIBLER, Gustavo - Design of an auxiliary electricity supply system from photovoltaic panels for residential use. Centre for Exact and Technological Sciences (CETEC) of the UNIVATES University Centre - 2015.

■ SEBASTIAO, Antonio Pedro - Mozambique's electrification model: its importance for development, international business and economics, 2016, e-issn 2183-3265, http://www.cigest.ensinus.pt/pt/edicoes.html.

■ SOLAX BUNDU POWER - Reseller of Solar Materials (inverter and solar panel), South African factory, January 2020.

■ SOUSA, Filipe Alexandre de; OLIVEIRA, Manuel A. S. de - Photovoltaic Solar Energy Installer Technician Course. 2015.

■ UNESCO (United Nations Educational, Scientific and Cultural Organisation); 2012 - International Year of Sustainable Energy for All; UNESCO Association Schools; 2012.

■ UNIVERSIDADE ZAMBEZE - Rules for the Presentation of Scientific Work - Research Projects, Dissertations and Theses - UZ DP DC 2012.

■ WHITE, Frank M "Fluid_Mechanics_5th_Ed, America 2001".

Annexes

Annex I - Mapping of the Gorongosa district and study site, Sofala

CHEMBA
MARINGUE
CAIA
GORONGOSA
CHERINGOMA
MARROMEU
MUANZA
NHAMATANDA
DONDO
CIDADE DA BEIRA
BUZI
CHIBABAVA
MACHANGA
Zona fora da reserva do PNG
MARINGUE
GORONGOSA
MUANZA

Annex I.1 - Mapping of regions outside the PNG reserve and study site, Gorongosa - Sofala

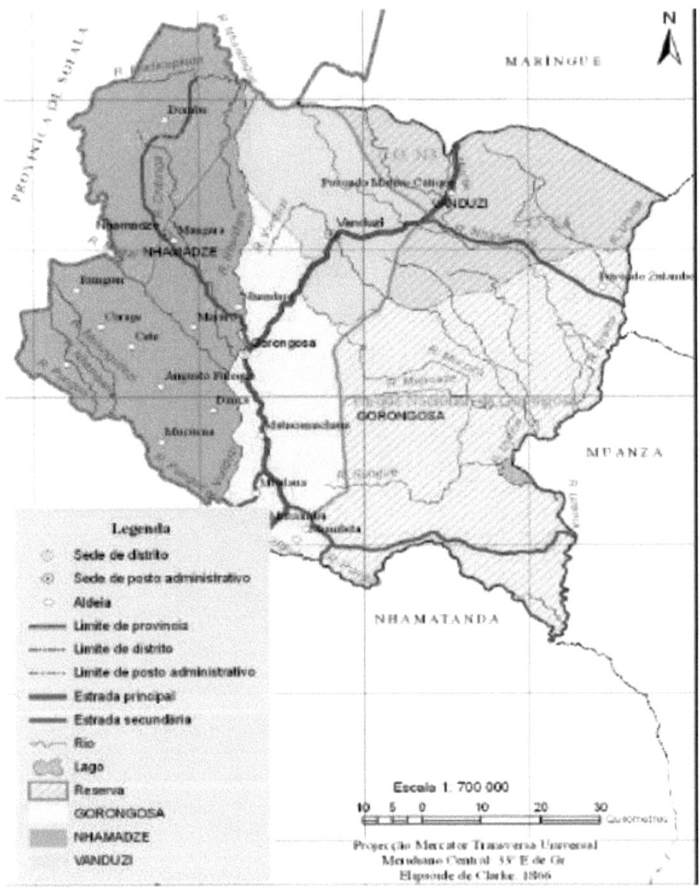

Annex II - Monthly energy consumption for each consumer in the Mateus Catique community, Gorongosa

Equipment	Power [W]	Quantity [u]	Hours/day	Daily consumption [kW/day]	Monthly consumption [kWh]
Starter Charge (Household Essentials)					
Computer	300	2	5	3	90
Monitor	12	2	5	0	4
Television	90	2	8	1	43
DVD/Decoder	45	2	8	1	22
Laptop	25	2	5	0	8
Simple amplifier	250	1	8	2	60
Electric pump	400	2	4	3	96
Lamps	30	8	8	2	58
Additional load forecast					
Simple fridge	250	1	24	6	180
Iron	200	1	3	1	18
Kettle	150	1	8	1	36
Total				**20**	**602**

Source: Author (2020).

Annex III - Photovoltaic system - General connection

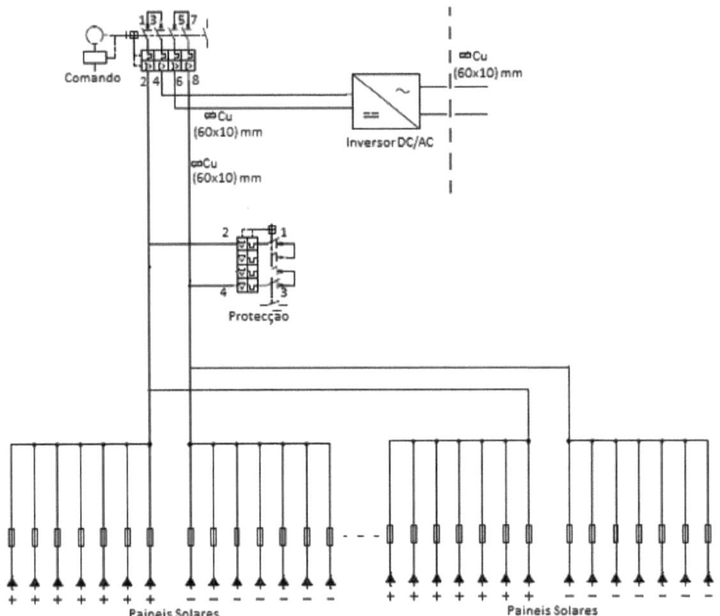

Annex III. 1 - General Connection - for Network or customers

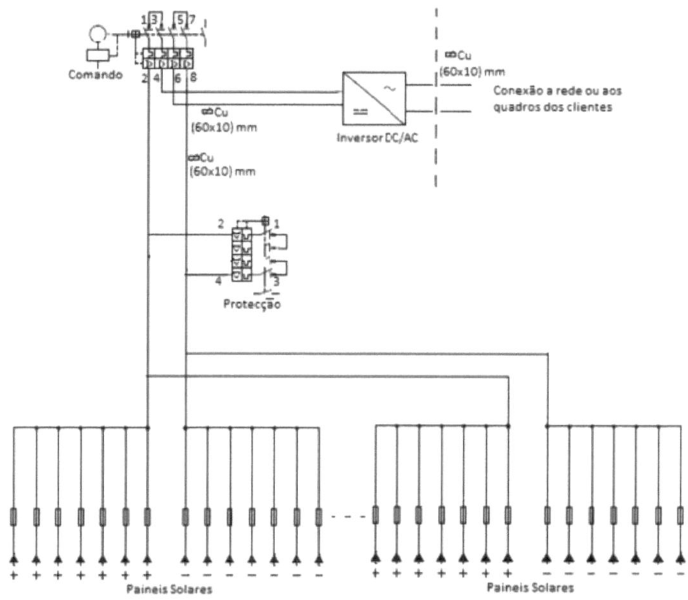

Annex IV - Switchboard connection for customers (residents)

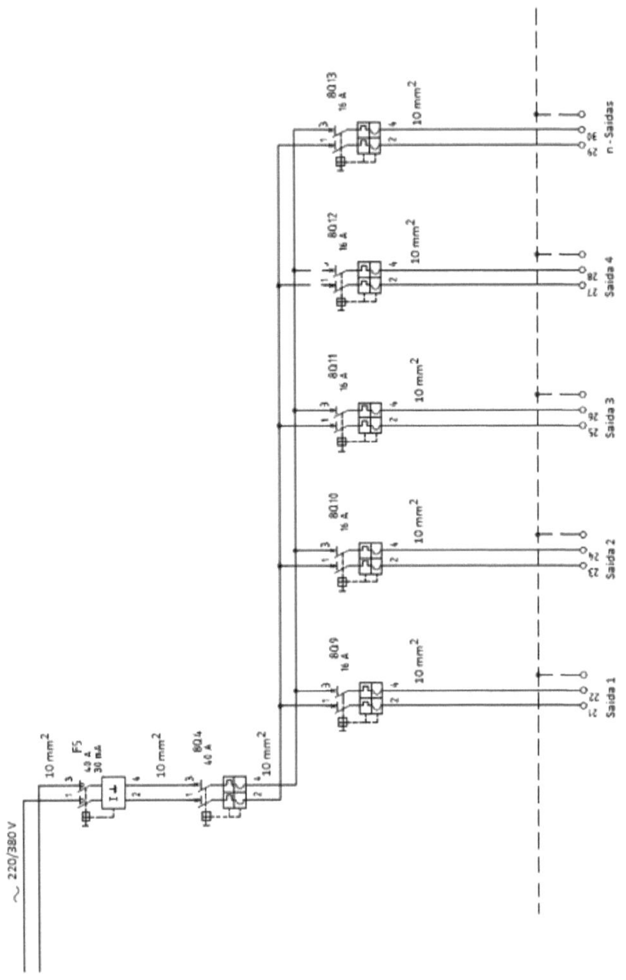

Annex V - Field of application of hydraulic turbines

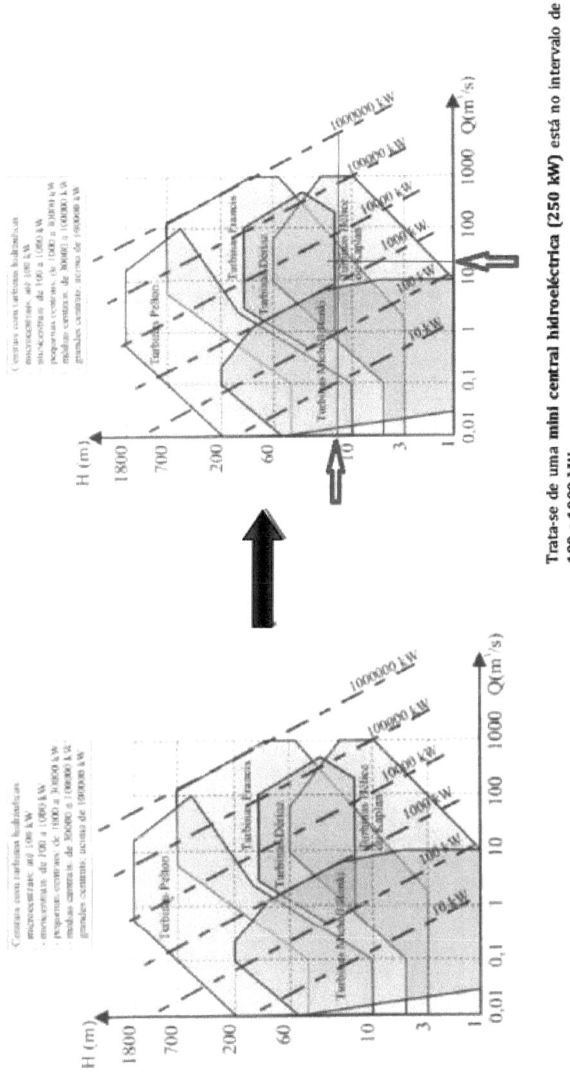

This is a **mini hydroelectric power station (250 kW)** in the **100 to 1000 kW** range.

Annex VI - Energy Estimates and Electrical Characteristics

Table A5 - Estimated daily energy consumption

Equipment	Power [W]	Quantity [u]	Hours / day	Daily consumption [kW/day]
		Initial Charge (Household Essentials)		
Computer	300	190	8	456
Radio	40	195	5	39
Television	90	300	10	270
DVD/ Decoder	45	250	10	113
Laptop	25	100	7	18
Landline	15	60	10	9
Electric pump	400	98	4	157
Lamps	30	800	8	192
Total				**1196**

Source: Author(2020)

Table A6 - Electrical characteristics of the Perkins generator, model 1306A-E87TAG3 and with a power of
power of 253kVA

Perkins 250kVA generator	
Type of chain	Three-phase
Operating voltage range	200 V-480 V
Frequency	60 Hz
Speed	1200 rpm
Emergency power (STPJ) (kVA/kWJ	253.3 kVA / 202.6 kW
Continuous power (PRPJ) (kVA/kWJ	231.4 kVA/185.1 kW
Autonomy 75% (HJ	15.4

Source: Alibaba.com consulted on 15/11/2020 at 10:53 am

Printed by Books on Demand GmbH, Norderstedt / Germany